Economics and Public Policy

The MIT Press
Cambridge, Massachusetts, and London, England

Economics and Public Policy:

The Automobile Pollution Case

Donald N. Dewees

This book was set in Linotype Baskerville
by Arrow Composition, Inc.,
printed on Fernwood Opaque
and bound in G.S.B. S/535/83 "Lime"
by The Colonial Press Inc.
in the United States of America.

Library of Congress Cataloging in Publication Data

Dewees, Donald N
    Economics and public policy.

    A version of the author's thesis.
    Includes bibliographical references.
    1. Automobiles—Environmental aspects—United States. 2. Air—Pollution—Economic aspects—United States. 3. Environmental policy—United States. 4. Automobile exhaust gas. I. Title.
TD886.5.D48   363   74-1239
ISBN 0-262-04043-3

Preface   vii

Preface

For a variety of reasons, automobile pollution control has been a major
source of public concern over the last several years in the United States and
in other countries. While it is justly identified as a serious problem with a
legitimate demand on the nation's resources, it is not the first problem of
its kind to arise. Until the 1950s, the coal-burning railroad locomotive was
one of the largest sources of air pollution in many cities and towns, where
heavy soot covered everything on the wrong side of the tracks for a consider-
able distance downwind. The horse-drawn carriage resulted in a pollution
of its own, which must have created aesthetic problems, particularly on hot
and humid summer days. Even the oldest lithographs of riverboats steaming
along the Mississippi and other arteries of commerce show billowing clouds
of smoke, which must have caused more frequent bathing by passengers and
inhabitants of the downwind shores than the delicate sensibilities of the
day would otherwise have demanded.

The internal combustion engine may be replaced by some other power
plant that does not cause the same air pollution problems. History suggests,
however, that even a radically different motor would probably have undesirable side effects of its own, perhaps not perceived or understood at its
introduction. While such side effects might be far preferable to the exhaust
of the internal combustion engine, they would nonetheless have to be dealt
with. In short, we might greatly reduce current pollution levels, but we are
most unlikely to develop a truly pollution-free vehicle, however close we
could approach that ideal. It is, therefore, necessary to decide how much
and what kinds of pollution we are willing to tolerate in order to enjoy the
undoubted benefits of mobility produced by the automobile. This is a problem in public policy that will endure long after current engines have been
superseded. Analysis of this kind of question requires some consideration of
economic issues.

It is the purpose of this study to bring together the tools and concepts of
economic analysis with available technological, medical, economic, and
other data in a framework that can organize these data to suggest the relative merits of alternative solutions to the policy problem. Detailed statistical
analysis has been relegated to appendixes so that the main body of the work
will be intelligible to those whose training or interest does not extend to
multivariate regression analysis and hedonic pricing techniques. Familiar-
ity with basic economic concepts has been assumed throughout, but this

is probably not necessary to understand at least the conclusions of the study, if not all the steps taken in reaching them. No one city or airshed has been selected as a focus, because it is felt that while the problems of Los Angeles, New York, and Toronto may be quite different, they share many common elements that can be treated in a general way. The policy recommendations are defined for areas with problems of different degrees of severity. It requires only a single additional step to decide how severe the problem is in a particular city or state, and to design the suggested policies specifically for that area.

The research reported on here was conducted at Harvard University under a Health Economics Fellowship granted by the Carnegie Corporation of New York, with assistance from the Environmental Systems Program of Harvard University, funded, in turn, by the National Science Foundation. Martin Feldstein, Mahlon Straszheim, and Henry Jacoby contributed greatly to the formulation of the study and its execution. Zvi Griliches advised on some of the econometrics. The work continued at the University of Toronto, with support from the Institute of Environmental Sciences and Engineering for the research and the production of the manuscript. A number of people made helpful comments, including Frank Hooper, Richard Caves, and two reviewers. Christine Purden edited and rearranged the manuscript, Bill Sims made some calculations, and Jessie Leger typed with a skilled hand. Nancy Reynolds prepared the earlier thesis version. I owe a large debt to these people and to others whose contributions are not enumerated but are still important.

Toronto, 1973

Economics and Public Policy

Introduction

1

The Problem

Automobile pollution control is a public policy problem that occupies a
large and growing place in the national economy. New cars built in 1973
will cost their owners an extra $2 to $4 billion over their lifetime because
of pollution controls; and, by 1975, this figure could jump to $6 to $9 billion. Thus, we may soon spend more on controlling auto emissions than the
federal government spends on new highways, one of the gigantic public investment projects in modern times. Many pollution controls increase fuel
consumption just as gasoline prices have begun to rise after years of
stability.

These expenditures are large because the problem is serious. For more
than a century, transportation activity has been a substantial contributor to
air pollution in America. From the early river steamers, pouring smoke
from their wood- or coal-fired boilers, through the steam railroads that
showered smoke and ashes along every track in the country, to the automobile, with its emission of invisible and sometimes deadly gases, the blessing
of greater mobility has always brought the curse of foreign matter in the air.

Currently, about 60 percent (by weight) of all air pollution in the United
States is emitted by motor vehicles, and between 30 and 60 percent of the
total weight of the three main pollutants—hydrocarbons, carbon monoxide,
and nitrogen oxides (almost 90 percent of transportation emissions)—
comes from automobiles. (See Table 1.1.) Indeed, automobiles produce

Table 1.1. Sources of Selected Air Pollutants (1968)

Source	HC (tons/yr)	(%)	CO (tons/yr)	(%)	NO_x (tons/yr)	(%)
Motor Vehicles						
Gasoline	15.2	47.5	59.0	58.9	6.6	31.9
Diesel	0.4	1.3	0.2	0.2	0.6	2.9
Aircraft	0.3	0.9	2.4	2.4	—	—
Railroads	0.3	0.9	0.1	0.1	0.4	1.9
Ships	0.1	0.3	0.3	0.3	0.3	1.4
All Other	15.7	49.1	38.1	38.1	12.8	61.8
Total	32.0		100.1		20.7	

Source: U.S. Department of Health, Education and Welfare, Public Health Service, *Control Techniques for Carbon Monoxide, Nitrogen Oxide, and Hydrocarbon Emissions
from Mobile Sources* (Washington, D.C.: U.S. Government Printing Office, 1970), Table
2–3, p. 2–15.

many times more pollution than all other modes of transportation combined. While buses, trucks, trains, and aircraft can cause serious problems in areas where they are heavily concentrated, their aggregate contribution to pollution is small (only a few percent); and the pollutants emitted by diesel and jet engines are not as harmful as those produced by gasoline engines.

The known and suspected ill effects of automotive pollutants have been a cause of public concern for over a decade.[1] In heavy traffic, carbon monoxide emissions can become so concentrated that they may affect driver performance and cause headaches among motorists and traffic policemen. Automotive emission of lead compounds may be a source of subclinical injury. Hydrocarbons and oxides of nitrogen may be harmful directly; but, more important, they combine under some conditions to form photochemical smog, which turns the atmosphere a murky brown, irritates the eyes, nose, and respiratory system, causes respiratory illness, and is even harmful to many plants and materials.[2]

Current population and vehicle use trends continually increase the potential harm from automotive emissions. If the number of registered automobiles continues to grow at the present rate of 5 percent per year, almost four times as many will be registered by the year 2000 as in 1972. Just to maintain present air quality, emission rates will have to be reduced to 25 percent of the average of the 1972 fleet. Any improvement in the air quality will, of course, require even further reduction.

Extensive programs for control of automobile emissions are already under way. Since 1960, the California Motor Vehicle Pollution Control Board has been responsible for issuing certificates of approval for motor vehicle pollution control devices that have been demonstrated to be reasonably effective at moderate cost. When two or more devices of a given type are approved, they become mandatory on new cars sold one year later in the state of California. At the federal level, the Administrator of the Environmental Protection Agency is charged with setting emission standards for new cars and certifying the compliance of those that meet the standard.

Table 1.2 shows the standards that have been imposed on new cars to

1. These are described in detail in Appendix A.
2. In the Los Angeles basin, photochemical smog causes such irritation on over 30 percent of all days, affecting up to 10 million inhabitants of the area. See Chapter 3.

date and those now enacted for application through 1976. Hydrocarbon and carbon monoxide emissions were first regulated in California in 1966, and for the entire United States in 1968. At that time, they were the only automotive pollutants for which abatement technology was available; and ironically, through the imposition of these controls, emissions of the oxides of nitrogen were increased. The latter are now regulated by federal laws effective in 1973. Clearly, the standards for 1975–1976, if they can be met, will represent great reductions from uncontrolled emission rates. This qualification is important because it has been suggested that these standards cannot be met at all, or that the cost of meeting them will be prohibitive.

The automobile occupies such a large part of the economic activity of the country that a significant percentage increase in motoring costs means a significant increase in the cost of living, and perhaps significant changes in the industry itself. In 1968, more than 83 million automobiles were registered, worth over $200 billion when new; new car purchases totaled $23.5 billion. Automobile vehicle miles in 1968 were over 800 billion, and have continued to rise at a rate of 5 percent per year for over a decade. Total automobile-related expenditures in the same year exceeded $90 billion;[3] and in 1965, spending on automobiles by households constituted 13.1 percent of personal consumption expenditures.

In 1972, one automobile manufacturer's procedural errors invalidated the tests that would certify one of its engines as meeting the 1973 emissions standards and thereby permit its production in 1973. When it appeared that repeating the test might postpone production of some 1973 models, the shudder that passed through Detroit passed similarly through Washington, transmitted by the specter of men thrown out of work for a few months because of pollution controls. Because production was not, in fact, seriously interrupted, a stark confrontation was narrowly avoided between the values placed on further automotive pollution control and bread-and-butter economic issues such as unemployment. It would have been an interesting confrontation.

---

3. Automobile Manufacturers Association, Inc., *1970 Automobile Facts and Figures* (Detroit, Mich., 1970).

Table 1.2. New Car Emission Standards

Year	Crankcase HC	Evaporation HC	Exhaust HC	CO	NO$_x$
Uncontrolled		2.77 gm/mile	10.2 gm/mile (834 ppm)	76.9 gm/mile (3.3%)	4.0 gm/mile (1,000 ppm)
California 1963	0.15% a				
California 1966	0.10% a		275 ppm b	1.5% b	
U.S. 1968	0		275 ppm b	1.5% b	
U.S. 1970 e	0		2.2 gm/mile	23 gm/mile	
U.S. 1970 d	0		4.6 gm/mile	47 gm/mile	
U.S. 1971 e	0	6 gm/test	2.2 gm/mile	23 gm/mile	
U.S. 1971 d	0	6 gm/test	4.6 gm/mile	47 gm/mile	
U.S. 1972 e	0	2 gm/test	3.4 gm/mile	39 gm/mile	
U.S. 1973 e	0	2 gm/test	3.4 gm/mile	39 gm/mile	3.0 gm/mile
U.S. 1975 e (original) f	0	2 gm/test	0.41 gm/mile	3.4 gm/mile	3.1 gm/mile
U.S. 1976 e (original) f	0	2 gm/test	0.41 gm/mile	3.4 gm/mile	0.4 gm/mile
U.S. 1975 e (interim) f	0	2 gm/test	1.5 gm/mile	15 gm/mile	3.1 gm/mile

Sources: For California 1963 and 1966, and U.S. 1968: Department of Health, Education and Welfare, Public Health Service, *Control Techniques for Carbon Monoxide, Nitrogen Oxide, and Hydrocarbon Emissions from Mobile Sources* (Washington, D.C.: U.S. Government Printing Office, 1970), p. 3–5.

For U.S. 1970 and 1971: *The Economics of Clean Air*, Report of the Administrator of the Environmental Protection Agency to Congress, March 1971, U.S. Senate Doc. No. 92–6, p. 3–8.

For U.S. 1972, 1973, 1975, and 1976: *Federal Register* 36:228, Nov. 25, 1971, p. 22452, and *Federal Register* 36:128, July 2, 1971, p. 12652.

a. Emissions as a percentage of fuel supplied.

b. For engines with placement exceeding 140 cu in. For smaller engines, the standards are:

Displacement	Hydrocarbons	Carbon Monoxide
50–100 cu in.	410 ppm	2.3%
101–140 cu in.	350 ppm	2.0%

Here, the hydrocarbon concentration is measured in parts per million (ppm) by volume.

c. Standards as originally published using 1970 test procedures.

d. Equivalent standards using 1972 constant volume sampling (CVS) test procedures.

e. CVS test procedures.

f. In April 1973, the original 1975 standards were postponed one year, and interim standards adopted for 1975.

Scope of Study

The literature on automobile pollution control is vast, and growing at an ever-increasing rate. Much is written on the technology of pollution control, from investigation of further modifications to the internal combustion engine to speculation on the possibilities for cars powered by fuel cells and atomic energy. The medical literature in which the effects of these pollutants are examined on man and other creatures is similarly extensive. In some studies, researchers have examined the costs of achieving various levels of control. So far, however, there has been no thorough economic study of this multibillion-dollar problem. There has been no comprehensive treatment of the problem that draws these elements together in a framework within which quantitative analysis of policy choices can show the relative advantages of alternative policies. The study presented here offers such a framework.

There are three key questions to be considered in pollution control policy formulation. It must be decided, first, how much pollution can be allowed; second, what technologies should be used to reduce pollution; and third, what administrative methods are best for achieving the desired improvement. These questions are closely interrelated; technological capability will usually influence the desired pollution level, and it also may affect the type of policy to be adopted.

In this study, we will develop a methodology for answering these questions and explore the policy implications for the analysis thus performed. This requires an examination of the benefits that accrue from pollution control programs. It also requires an examination of specific abatement devices to determine their relative cost effectiveness in reducing auto emissions. And finally, it requires assessment of strategies to cause the utilization of available technology, or development of new technology, to judge which will best achieve the desired result.

In considering the technological aspects of pollution control, recommendation of one abatement device over another is subject to the probability that technological innovations will continue to appear, and the best device of today may be obsolete tomorrow. Thus, the development of a method for evaluating devices is as important a contribution as the actual answers reached here. Our focus will be on devices that may be employed over the short run—five years or less, since the technology that could be in production over the next five years already exists, if only in laboratory models and prototypes. Any devices that might appear after this period either are not well defined at present or could have costs and performance so different from current technology that any analysis made on existing data would be meaningless.

In selecting an abatement strategy, however, we will consider long-run conditions, for a period of ten years or more. The policies for control, once established, are not likely to be soon changed, particularly if they significantly reduce pollution levels and thus public concern. The term *strategy*, as used here, means the complete set of government programs focusing on the problem of air pollution caused by automobiles. An important aspect of policy implementation in this area is that, since the production and operation of automobiles takes place in the private sector, the setting of pollution control standards inevitably implies government involvement in private decision making. Thus the choice of appropriate policy instruments is a question as difficult as selecting the proper degree of pollution control.

Plan of Study

In the next chapter, the criteria are developed for selecting among various pollution control systems, programs, and strategies. Abatement alternatives are discussed and clarified. Chapter 3 examines in detail the benefits to be derived from reducing automobile emissions, focusing on the resultant improvements to health. In addition, several indices are developed to measure pollution abatement.

Chapter 4 presents the models and assumptions used to determine the cost of abatement programs. Equations are developed for computing each of several cost elements.

Chapter 5 explores the relationship between gross vehicle parameters and fuel consumption, and between fuel consumption, fuel composition, and pollution production. The analysis of fuel consumption is useful not only for its pollution implications but also for evaluating policies aimed at reducing fuel consumption to adapt to the increases in gasoline prices that may occur during the 1970s because of higher royalties to oil-producing nations and increasing marginal costs of finding petroleum and extracting it from the ground.

Chapter 6 reviews the evidence on the demand for motoring, in order to assess the feasibility of reducing pollution by curtailing motoring itself. The implicit cost of such reductions is estimated.

In Chapter 7, the cost of reducing emissions per vehicle-mile is studied through a review of twelve different abatement devices, either in use already or proposed for the future. The cost and effectiveness of each is computed, and a marginal abatement-cost curve derived.

In Chapter 8, all our results are combined and evaluated. Several feasible technical means of abatement are identified, and some estimates are made of the desirable degree of abatement. Then an overall strategy for automobile pollution control is developed, consisting of several programs at different government levels. The study concludes with criticism of past policies and a series of recommendations for government action.

Policy Evaluation: Criteria and Alternatives

2

Pollution Control Objectives

Public policy for pollution control may pursue a variety of objectives. It has been suggested that pollution should be completely eliminated by banning all noxious emissions. In an industrial urban society, this is not a practical alternative, since it would mean abandoning almost all economic activity. Some have suggested that abatement should proceed as far as possible— that is, that the best available technology should be installed, regardless of cost. Others have proposed that pollution should be reduced to the point where it has no impact on present ecological conditions or inflicts no long-run damage on major life systems.

The last objective is one with which few would quarrel. The problem is that current information is not sufficient for us to evaluate such long-range concerns. If scientists could say with some certainty what the impact of different rates of emission might be on climate, living systems, and man, such environmental impacts could be evaluated and objectives set accordingly. In most cases, however, this is impossible; it is generally feared that many years from now new and harmful effects of pollution will emerge that can not be anticipated at present. In the light of this concern, there can be no doubt about the need to control pollutant emissions to some degree.

In theory, at least, economics offers an objective for determining the proper degree of pollution control. To begin with, consider the automobile as a durable consumer good. It provides services to consumers in the form of vehicle-miles or passenger-miles of transportation, and concurrently it produces a second good: air pollution. Air pollution is an unpaid economic factor in that the motorist does not pay those who suffer from his pollution for his use (and degradation) of the air. This is a classic example of a technological externality, where the production of one good has an economic impact on others without there being any market transaction.[1]

---

1. For analysis of the problem of air pollution in terms of externality theory, see Francis M. Bator, "The Anatomy of Market Failure," *Quarterly Journal of Economics,* 72, no. 3 (August 1958): 351–379; R. M. Coase, "The Problem of Social Cost," *Journal of Law and Economics,* 3 (October 1960): 1–44; Otto A. Davis and Morton I. Kamien, "Externalities, Information, and Alternative Collective Action," in Robert Dorfman and Nancy S. Dorfman, eds., *Economics of the Environment* (New York: W. W. Norton and Co., Inc., 1972), pp. 69–87; Allen V. Kneese, "Air Pollution—General Background and Some Economic Aspects," in Harold Wolozin, ed., *The Economics of Air Pollution* (New York: W. W. Norton and Co., Inc., 1966), pp. 23–39; and J. E. Meade, "External Economies and Diseconomies in a Competitive Situation," *Economic Journal,* 62 (March 1952): 54–67.

The existence of this externality has two consequences: First, since the motorist causes a social cost which he does not pay for, he produces more pollution than is optimal. The individual who uses a resource that is free tends to use more of it than he would if he paid for it; he would use less also if a market existed in which those who suffered from the pollution could pay him to produce less. In either case, the price paid would establish the marginal social cost of pollution and would thereby reduce it. Second, if motoring and air pollution are complementary in production and more air pollution is produced than is optimal, provided that no other market imperfections exist, more motoring than is optimal will result. Any mechanism that is used to control pollution should also reduce motoring by some amount.

In the long run, the joint production function for motoring and air pollution is characterized by variable rather than fixed proportions. Depending on relative prices, air pollution per mile may vary upward or downward, since a variety of technology is available that can reduce pollution per mile at increasing marginal cost. Thus a program that reduces air pollution need not reduce motoring by the same percentage. The reduction in motoring will depend upon the shape of the joint production function cost curve and the degree of pollution reduction.

The economically efficient solution to a technological external diseconomy such as automobile pollution is to control emissions until the marginal cost of abatement is just equal to the marginal benefits of abatement. If it were possible to estimate both costs and benefits accurately, choice of the proper degree of abatement would be relatively straightforward. In the case of the automobile, which is typical in this regard, it is particularly difficult to estimate benefits, so this is not a very useful basis for choosing the proper emission rate. It is, however, still desirable and will be pursued to the extent possible.

It is important to note that this focus on economic efficiency need not be inconsistent with the goal of preserving existing life systems (or other environmental objectives), provided that the environmentalist's value system is adopted in the measurement of the benefits of alternatives. The only problem is that there will be a tendency, when benefits are measured, to count those that can be quantified and ignore the rest, while actually some value should be assigned to the rest, even if it must be done subjectively.

In addition to deciding how much pollution should be allowed, it is necessary to decide how the abatement should be achieved. Here it is assumed that the lowest-cost means of abatement is the best. Accordingly, whatever the objective of the control program, efficiency in achieving that objective will always be pursued. This provides a method for evaluating alternative means of achieving a degree of abatement, even if there is no immediate agreement on the actual abatement goal.

A theoretically attractive way to equate marginal abatement costs to marginal benefits would be for motorists to pay the marginal social cost of their pollution for every mile driven.[2] Exactly what this cost would be would depend on the local air pollution level, population density, proximity of sensitive flora, and many other factors; thus it would vary for every mile driven. The motorist, anticipating this added motoring cost related to pollution, would select a vehicle in which the marginal cost of abatement was just equal to the average marginal social cost incurred in its use. He would also drive somewhat less than otherwise, because the marginal cost of abatement would increase the marginal cost of motoring.

Because the harm from automotive pollution is not a commodity traded in a market, we cannot observe directly a value for the marginal social cost of this pollution in every area of the country and at all points in time. Equally important, the technological problems of determining where each mile was driven by each automobile in the country—so that the proper charge could be applied—are immense. Thus while the government may limit automotive emissions and even increase the cost of motoring, only rough approximations to the optimal policy are likely to be achieved.

There are two ways to test the government's chosen course of action, to see how closely it approximates the optimal solution. First, it should result in a pricing scheme that imposes on motorists the marginal social cost of their use of automobiles. Given that in the optimal situation marginal social benefit equals marginal social cost, alternative strategies can be ranked by the degree to which their prices approach the optimum.

The second test recognizes that a solution is optimal, not just because the prices are right, but because those prices cause the proper amounts of all

2. For a discussion of alternative solutions, see Kneese, "Air Pollution," and Allen V. Kneese, Robert V. Ayres, and Ralph C. d'Arge, *Economics and the Environment* (Baltimore: Resources for the Future, Inc., Johns Hopkins Press, 1970).

goods, including the two joint goods of primary interest, to be produced. Thus the selected strategy should result in the production of optimal amounts of pollution and motoring, regardless of the price system in existence. If the same effect can be achieved by constraints or other mechanisms, the solution deviates from the optimal only in the distribution of income.

Tools for Evaluation

Given that the objective is to produce a public good—clean air—in such a manner that the marginal costs of production equal the marginal benefits, the different strategies for achieving this objective may be evaluated through cost-benefit analysis.[3] This tool is commonly used to evaluate investments made by government, but it can be applied equally to the evaluation of investments in goods that are produced in the private sector by government order. Alternative programs can be compared in terms of the costs of the various technologies that achieve different levels of pollution abatement and the respective social benefits that accrue from reduced air pollution levels.

When we evaluate various strategies by cost-benefit analysis, we must remember that in the real world many factors may prevent achievement of the optimal solution. Any abatement program probably will be imposed at a fairly high level of aggregation—that is, standards will be set on a national scale. Abatement costs will vary greatly among vehicles and pollution levels, however, and the marginal social cost of a given unit of pollution will range widely among different parts of the country, so that any uniform pollution standard or technology will not be entirely optimal. For example, if marginal social cost and benefit are calculated in terms of a national average, estimated abatement costs will be the same for the occasional motorist in a rural community as for a regular commuter in New York.

The discussion thus far has been couched in terms of economic efficiency, without regard to income-distribution consequences. While this is the traditional framework for most microeconomic analysis, it is worthwhile exam-

3. For an example of the means to determine the demand for environmental quality, see P. Bohm, "An Approach to the Problem of Estimating Demand for Public Goods," *Swedish Journal of Economics,* 73, no. 1 (March 1971): 55–66; and P. Davidson, "The Valuation of Public Goods," in Dorfman and Dorfman, eds., *Economics of the Environment,* pp. 345–355.

ining the income-distribution consequences of pollution control policies even if they are not formally incorporated in the analysis.

There are those who suggest or demand that the auto manufacturers bear the cost of cleaning up their cars, presumably because of some accumulated guilt from the dirty old days. Economically, however, this makes little sense. Pollution controls will be perceived in the first place as a cost increase, which without price increases would reduce profits. The precise costs, however, are unknown except to the auto manufacturers (and maybe not even to them), and, with price changes every year for other reasons, it would be impossible to prevent passing this price increase along to the customer. If profits are to be regulated or limited, this can be done entirely independently of pollution control policies. Thus the most realistic assumption is that any costs would be borne directly by the motorist.

In a cost-benefit analysis, or any other public program evaluation that affects incomes of individuals, it would be desirable to have a model showing the income effect on persons of every income level and to have a social welfare function that evaluated each type of impact. While formal models are not available, some evaluation can be made. If motoring costs are increased by a uniform percentage for all motorists, the incidence of this increase will be regressive, like that of any sales tax, since automotive expenditures rise less rapidly than income. If the percentage increase in the cost of economy cars is greater than that of luxury cars, as seems quite probable for a given degree of pollution control, the impact will be even more regressive. If certain kinds of automobile trips—for example, to downtown areas—are restricted, made more expensive, or prohibited, the impact will depend upon the income classes that tend most to make that kind of trip. Here it will generally be assumed that the impact of abatement programs will be regressive to the same degree as a sales tax; where this is not the case, the distributional effects will be discussed specifically.

Other authors have noted that environmental quality tends to be a luxury good that is desired more by high-income persons either because tastes are different or because it ranks after food, shelter, and other necessities on most consumers' list of priorities. Since it is a public good that will be consumed in similar proportions by most residents of any given metropolitan area, the regressive incidence of its costs are mirrored by the progressive enjoyment of its benefits. This exacerbates the distributional inequities of

pollution control programs generally. Where alternative policies may alter
the incidence of benefits by income class, this will also be noted as affecting
the distributional effects of the program.

Since we are concerned here with the policies that regulatory bodies
might be required to implement, we should consider the functioning of
bodies that presently have or will assume pollution control responsibility.
Here we look to other regulated industries for our models—for example,
public utilities, insurance, communications, and the transportation industry. The common pattern at both the state and the federal level is for a
newly established regulatory body to be energetic and thorough at first but
to relax as time passes. There is a tendency in the long run for the regulatory body to protect and promote the industry as much as it does the public
interest. This reduction of independent motivation results in part from the
fact that the same men are working together over a long period of time and
in part from the frequent movement of personnel between the industry and
the regulatory body. An individual who was employed initially in the industry and who is temporarily active on the regulatory body is not likely to
jeopardize his chances of returning to the industry by pursuing some line
of policy that will alienate his former employer. Thus it should be recognized that a body empowered to regulate automobile air pollution may
not always regard the public interest as its first concern. This suggests the
desirability of legislative mandates that do not leave entirely to the particular administrative body the determination of how rapidly abatement
programs should proceed.[4]

It is important to note that pollution control is like any public good in
that everybody benefits a little, but no individual benefits enough to have
a strong interest in it. Without a powerful lobby behind pollution control,
regulatory bodies tend to be chronically underfunded, as many state agencies have been in the past. State or local automobile pollution control agencies, and to a lesser extent the federal government, are therefore unlikely to
have adequate numbers of highly skilled technical personnel capable of
analyzing complex technological issues or of conducting extensive measurement and surveillance programs.

Because of aggregation problems, income distribution, and the com-

4. For an analysis of the effectiveness of regulatory agencies, see Paul W. MacAvoy, ed.,
The Crisis of the Regulatory Commissions (New York: W. W. Norton and Co., Inc., 1970).

plexity of the solutions to be considered, it is not possible to set up a single measure or objective function against which all strategies can be evaluated. Rather, several criteria must be used, based on economic theory and practical considerations. First, a strategy is preferred to the extent that it tends to impose upon the motorist the marginal social cost of his motoring activities. Second, and closely related to the first, preferred strategies are those that produce benefits at the lowest possible total cost. Third, and independent of the first two, a strategy is preferred to the extent that it tends to increase the rate of technological progress in automobile pollution control. The production function for innovation or technical progress is not well defined or easy to quantify; but if one plan is more likely than others to produce rapid progress, it is clearly to be preferred, other things being equal or nearly so. Fourth, the regulatory body should require the least possible amount of technical expertise, since the acquisition of such expertise can be expensive beyond the means of the usual government agency. Fifth, a strategy is preferred if it has a desirable rather than an undesirable impact on income distribution. It is assumed that, at a minimum, this means not increasing current income inequalities.

Generally, cost-benefit analyses produce a final figure of merit in the form of a cost-benefit ratio, a present value, or an internal rate of return. Because of the difficulty of evaluating the benefits of pollution reduction, we will evaluate alternative programs in terms of relative costs and effectiveness in this study. The figure of merit, then, is the cost per unit of reduction in pollution. Present value is not an appropriate measure in this instance, since benefits are expressed in physical rather than dollar terms.[5]

The benefits of abatement can be considered to be nonlinear and marginal benefit to increase with the density of pollution.[6] To the extent that this nonlinearity exists, projects cannot be compared in isolation; we must know at what level of abatement the project will operate, since its effects will be more valuable per unit at high than at low pollution densities. Fortunately, we can usually establish the situation to which a particular project or device may be applied, and therefore can determine the weighting of its effects.

Another problem will arise in comparing mutually exclusive projects of

5. Several indices for the measurement of costs and benefits are described in Chapter 3.
6. See Chapter 3 for a discussion of this assumption.

different magnitudes. If we considered cost effectiveness alone, we might choose a very efficient project that could achieve a 5 percent reduction and forgo a slightly less efficient project that would achieve a 20 percent reduction. A sensible evaluation of all projects necessarily requires the use of cost effectiveness but should combine all results with an indication of the total abatement that each project will achieve. Thus we should be able to identify both highly cost-effective projects *and* highly cost-effective projects that can achieve a given degree of abatement.

Classification of Alternatives

There are many ways to control automobile air pollution and many changes that controls might reasonably be expected to cause. To simplify our study, we have classified our alternatives along two dimensions—technological alternatives and administrative alternatives. Technological alternatives are the physical changes that directly cause a reduction in pollution emissions, from engine modifications to reductions in driving. Administrative alternatives are the policy instruments used to implement one or more technological alternatives, from rules and regulations to taxes. Within each of these alternatives, there are a number of divisions.[7]

TECHNOLOGICAL ALTERNATIVES

There are seven categories of possible technical changes. The specific alternatives considered here are those for which some data currently are available, but they do not preclude additional technological devices that may emerge in the future.

ENGINE DESIGN This can include all changes in the automobile power plant, from the addition of pollution control devices on existing engines, to the development of entirely new energy conversion systems. It can mean less pollution from new cars or less increase in pollution as the vehicle ages or both.

FUEL Some improvements in emissions may be obtained by changing fuels for current engines, modifying the hydrocarbon composition, or reducing the lead content. Other advances can come from a fuel change combined with an engine change—for example, lead-free gas permits the use of a catalytic muffler without lead fouling of the catalyst; the engine may be

7. Technological alternatives are discussed in detail in Chapters 5, 6, and 7; administrative alternatives are explored in Chapter 8.

converted to run on natural gas; or kerosene may be used in a steam-driven automobile.

ENGINE SIZE For any given size of automobile, a reduction in engine size can bring a reduction in fuel consumption per mile, until some low level of performance is reached. Such reductions in fuel consumption may affect pollution emissions.

VEHICLE SIZE For any given engine size, the amount of power needed to move the automobile decreases as vehicle size, particularly weight, decreases. Thus size reductions alone can reduce fuel consumption and, perhaps, emissions per vehicle-mile.

REGULATION OF MAINTENANCE Most people do not keep their automobiles in optimum running condition; thus their automobiles burn more gasoline per vehicle-mile and produce more pollution per gallon of gasoline than is necessary. Some gains, therefore, can be made through mandatory improvement of maintenance and modification of used cars.

REGULATION OF USE The most certain way to reduce total pollution is to reduce the total number of vehicle-miles driven. This can be done by some general restriction, such as rationing gas or raising its price, or by specific restrictions, such as limiting access to particular areas of the city or particular highways all or part of the time. An indirect method of regulation would be the provision of an attractive substitute for automobile transportation such as more accessible or cheaper mass transit.

ROADWAY MODIFICATION Some strategies may be used to redesign the roadway to minimize the impact of emissions on the surrounding area. For example, urban freeways may be located in areas that are not densely populated; or they may be constructed as subways, with tall stacks to disperse exhaust fumes far above the ground. On city streets, improved traffic control can produce a smoother flow of vehicles and less pollution per vehicle-mile. In the case of noise pollution, which is not directly considered here, shielding can be designed into the roadway to reduce noise levels somewhat in the immediate area.[8]

The items categorized above could be applied separately or could be

---

8. It has been suggested that the polluted atmosphere itself might be treated by filtering on a vast scale, or that it might be dispersed by giant fans. These schemes are likely to require such enormous amounts of power that we do not consider them to be reasonable alternatives for control.

combined in almost any way. An optimal control strategy might involve attacking the problem on several technological fronts.

The translation of the preceding list into a program for action would require identification of all the discrete alternatives within each category and the determination of the costs and effectiveness of each. Equally important, however, is consideration of the policy variables whereby one or more of the technological alternatives may be implemented. This problem of public regulation of private industry and persons will be considered in the next section.

ADMINISTRATIVE ALTERNATIVES

If public control is to be exerted over automobile pollution, some means must be found of injecting the government into industry decisions about vehicle design and manufacture and into private decisions about vehicle choice, operation, and maintenance. It must be decided which controls can best be exercised by federal, state, or local governments; how the public costs of the control program, if any, should be borne; and which administrative instruments are best suited to the implementation of the various technological alternatives. Above all, these decisions must be based upon a reasonable set of assumptions about the capability and effectiveness of the people and institutions that will be ultimately responsible for making a pollution control program work.

Both the state and the federal governments have the constitutional power to regulate motor vehicles and thus to take action with regard to pollution, although federal supremacy requires that the states not enact laws that are inconsistent with federal laws or regulations. Historically, state governments have found it easy to regulate motor vehicle owners through registration and licensing, insurance, and safety inspections, since both vehicle and owner are within the jurisdiction of the state, and dealers are subject to the usual state limitations on corporations. They have found it much harder to regulate manufacturers since, in most cases, the manufacturer is located outside the state and therefore does not come under its jurisdiction for many purposes. When the state's power over manufacture is limited to prohibition of the sale of nonconforming automobiles, it will be ineffective unless at least some of the manufacturers offer conforming vehicles for sale.

Local government is more restricted than state government in the types of regulation it can impose. Its primary powers are to regulate use of motor vehicles, impose taxes on vehicles and users, and approve the design of highway facilities.

Regulation by any governmental body may be divided into two general categories: restrictions that compel or prohibit certain actions; and pricing changes, such as taxes or subsidies, which act as an incentive to change. The following list of policy instruments deals with these in turn.

REGULATION OF MANUFACTURERS AND IMPORTERS Because there are very few manufacturers of automobiles, both domestic and foreign, regulations applying to them can be enforced relatively easily. Under such regulations, the manufacture and sale of noncomplying vehicles is likely to be prohibited with penalties ranging from fines to imprisonment. Regulations may specify particular equipment or specify the desired performance, so that manufacturers will constantly seek better and cheaper ways to achieve the performance objective.

REGULATION OF DEALERS AND SERVICE GARAGES It may be decided to regulate dealers rather than manufacturers, although there are thousands of dealers and only four domestic manufacturers. A restriction on the type of vehicle the dealer may sell, for example, will indirectly impinge on the manufacturer. Furthermore, if the dealer is made responsible for performance, he will be motivated to adjust new cars accurately to satisfy the specified design requirements. Used-car dealers may be required to adjust all the cars they sell to minimize their emissions. Service departments may be made responsible for the emissions from all the vehicles they work on. The last regulation may be imposed by either the federal or the state government, but the latter would be more likely to act at this level.

REGULATION OF OPERATORS AND OWNERS The most difficult requirements to enforce are those upon owners or operators themselves, since they number close to 100 million. Such regulations will indirectly affect vehicle design to the extent that people consider operating problems at the time of vehicle purchase. The most effective controls are those regulating the motorist's care of the vehicle and restricting its use.

REGULATION OF ROADWAYS Some changes can be made in the roadway itself to reduce noise reception and the local impact of gaseous pollutants. The federal or state governments may adopt standards of roadway design and require compliance with those standards as a condition for subsidization of highway construction. As local areas acquire more control over state and federal roads affecting them, they can at least informally influence the important design features. Government at any level can adopt standards with regard to traffic control.

TAX ON NEW VEHICLE SALES This is designed to alter the buyer's choice of vehicle and the manufacturer's vehicle designs in order to reduce pollution. It may be a graduated tax related to the projected lifetime emissions of the car, or it may be related to some gross vehicle parameter that correlates closely with emission rate, such as engine or vehicle size or the presence or absence of particular control devices. Because it operates at the time of purchase, it affects choice of vehicle and not subsequent operation of that vehicle. The tax may be levied by either the federal or state governments.

ANNUAL TAX ON VEHICLE This may take the form of a tax based on the design of the automobile, much like the excise tax just suggested, related to typical emissions or vehicle size. It will not affect operation, since it will be a fixed amount annually, but will have some effect on purchases as owners realize that they can cut their annual tax by buying an automobile with a lower pollution rate. Again, the tax may be imposed by either the federal or state governments.

VEHICLE USE TAX This tax may be levied upon every mile driven, through a regular reading of the odometer and the imposition of an appropriate charge for accumulated mileage; or it may be applied to one factor, such as gasoline purchases. The mileage tax will probably be imposed by the state, while either the federal or state governments may levy the gasoline tax.

TAX BASED ON EMISSIONS This is a form of effluent tax. The tax is based, not upon vehicle design or gasoline sales, but upon actual emissions for the year. The automobile may be inspected annually and the rate of emission recorded. That emission rate and the mileage driven during the year will become the basis for levying the tax. This tax will encourage people to drive less, to choose vehicles with lower emission characteristics, and to maintain their vehicles properly.

SUBSIDY OR IMPROVEMENT OF SUBSTITUTES All the alternatives discussed so far operate directly on the automobile. When the objective is to reduce total motoring, however, an effective program is to increase subsidies to urban mass transit, that is, to finance a reduction in fares or an improvement in service. Such a scheme may include the subsidization of intercity railroad, bus, and airplane systems and the financing of research and development for various forms of mass transit.

The tax alternatives suggested here all have their counterparts in the form of subsidies or tax rebates. If the marginal rate of the subsidy is the same as the tax, then the substitution incentive effect will be the same. The income effect, however, will be different; that is, a particular policy may have a regressive income redistribution effect if exercised in the form of a tax.

Economists generally tend to favor regulation through price variables if this will bring price closer to marginal cost (one of the conditions of achieving a Pareto optimum). But even if price strategies are decided to be the best control measure, there are still practical obstacles to be faced that are not applicable to legal constraints. It is difficult to convince many people that price changes are a practicable way to achieve quantity changes. In the case of pollution, where the real objective is to reduce an undesirable activity, many will argue that a price increase does not guarantee a reduction in pollution, because people may choose to pay it rather than comply. Worse, the rich will be able to continue polluting because they can afford the tax, while the poor will not. A constraint, in contrast, applies equally to rich and poor. This type of argument may reflect a welfare function that attaches no weight to the satisfaction of the rich or to the satisfaction of owning an automobile larger than bare necessity demands, or it may be irrational because of failure to understand the problem. In either case, it may be a substantial consideration in choosing a strategy.

Of the many categories for action listed above, both technological and administrative, some will not be formally analyzed here because of poor information or a strong a priori belief that they are likely to be dominated by some other choice. Of the technological alternatives, roadway modification will not be discussed. While in some cases this may be a highly cost-effective means of reducing pollution per vehicle-mile, particularly if it involves minimal new construction, no data are presently available to permit evaluation of the relative costs and benefits of such a scheme. It follows that we will also omit any discussion of regulations concerning highways, since such regulations apply specifically to roadway modification.

Another administrative instrument that will not be formally evaluated here is regulation of dealers and service garages. Because the owner of the vehicle is responsible for bringing it in for servicing, it seems unreasonable to expect the shop to take responsibility for the condition of a vehicle that may have been neglected for a long time. It is simpler and more logical to make the owner entirely responsible for the condition of his vehicle, perhaps by giving him a right of action against a service facility that fails to maintain the car as requested or as it purports to do.

Benefits of Pollution Abatement

3

In the formulation of any pollution control policy, the selection of an abatement objective must be based on knowledge of the benefits of abatement. Here we will look at the value to society of various reductions in the production of some specific air pollutants and the problems in determining this value.

Since air pollution, or pollution abatement, is a social good not normally bought in the marketplace, its value must be found elsewhere. One measure of this value is the economic loss caused by pollution, consisting of increased medical care, lost wages, physical discomfort, and property damage. Another measure is the value that individuals place on changes in perceived levels of pollution.

We will first select an index that will allow us to weight the relative importance of the four primary automotive pollutants; then we will explore the value of the harm done by these pollutants, converting the index to a measure of dollar loss per unit of pollution. Establishing the relative importance of the pollutants and their absolute value is a difficult empirical problem that is not finally resolved here, but in principle it is no different from any problem in public expenditure where the value to society of many kinds of social goods must be compared.

Losses Caused by Automobile Pollution

The recent concern over automobile pollution control is based on four factors: First and foremost, the public and the medical profession believe that this pollution is or may be harmful to the health of those who breathe it in regularly. Second, in many urban areas, automobile pollution is an aesthetic problem: it colors the atmosphere gray or brown, reduces visibility, has an offensive smell, and, in the case of photochemical smog, can be irritating to the eyes and the respiratory system. Third, this pollution, and particularly the products of photochemical smog, can be harmful to plants by reducing their growth or actually killing ornamental plants and trees and many agricultural crops. Finally, there is damage done to materials of all kinds. Automobile pollutants do not cause much visible soiling, but they can weaken some fabrics, and the ozone in photochemical smog weakens rubber, causing cracks in tires and other automotive components.

The construction of an economic loss function for automobile pollution, from which we could derive the benefits of various degrees of abatement, ideally would consist of quantifying the physical relationships in each of the four categories above. This would allow us to predict the physical or physiological impact of a given degree of abatement; then, if we assigned values to each type of physical impact, we could arrive at the dollar loss or benefit from a change in the pollution level. The function would have to be analyzed by specific pollutant, by area of the country, and by other environmental factors, so that abatement of individual pollutants in single states or cities could be evaluated.

Unfortunately, it is not possible to perform such a detailed and precise analysis. There are some data on the damage done in each of the four categories, but most of these data are not useful for determining the gross impact of a pollutant. For example, many laboratory studies have been conducted to measure the impact of a given pollutant on particular animals, plants, and materials. These may give some indication of the physical process involved in reaction to pollutants; but it is very doubtful whether there is value in applying results obtained in the artificial environment of the laboratory to conditions in the real world outside. Elsewhere, scattered estimates have been made of the magnitude of damage done by air pollution to plants, materials, and people in a given city or region; but these estimates usually treat air pollution as a homogeneous substance and cannot distinguish the effect of individual pollutants. The most frequently used proxies for air pollution are sulfur dioxide and particulates, which are easy to measure, but are entirely unrelated to automobile pollution.

Despite these shortcomings, the large amount of research that has been done on the subject does provide us with some useful information. In the pages that follow, we will examine some of the data regarding health effects of the automotive pollutants, as an example of what can be learned from the type of data that does exist in each of the four fields of impact. If an exact function cannot be generated, one might be able to determine the shape of the function and the limits on the total loss that may be suffered.

HEALTH EFFECTS OF AUTOMOBILE AIR POLLUTANTS

There are four main components of automobile pollution: carbon monoxide, lead, hydrocarbons, and nitrogen oxides. Several of these combine chemically to produce another atmospheric pollutant, photochemical smog. We will consider each of these in turn.

CARBON MONOXIDE Carbon monoxide (CO), is a colorless, odorless gas emitted with automobile exhaust at a rate of about 3 percent by volume if no control measures are applied. Urban concentrations range from two parts per million (ppm) to peaks of over 50 ppm.

The effect of carbon monoxide on humans arises from its affinity for hemoglobin, with which it combines in the bloodstream to form carboxyhemoglobin (COHb), reducing the capacity of the bloodstream to carry oxygen from the lungs to other parts of the body. When a person is exposed to air with a fixed percentage of carbon monoxide, his blood COHb will rise and asymptotically approach an equilibrium level for that atmospheric concentration. Since carbon monoxide is produced naturally by the body, the presence of small amounts of COHb in the blood is normal and not harmful.

The effects of carbon monoxide range from impairment of psychomotor function and reduced driving ability, through headaches and nausea, to death from very high concentrations. Schulte[1] found that errors in cognitive and psychomotor tests increased not linearly with concentration, but exponentially, in the form

$$y = 10^{(aX + b)} \ ,$$

where y represents errors, and X the COHb concentration. Concentrations of CO in automobiles and on congested highways reach levels where some of these effects may be expected in motorists and others exposed for long periods, such as traffic policemen.

While the long-term effects of exposure to ambient CO concentrations are unknown, in view of the known short-term effects, it appears that significant benefits would result from a reduction of emissions in heavy traffic areas. The benefits per unit of abatement either would be linear or would increase with increased density. It is not apparent whether there is a "safe" lower threshold level or not.

LEAD Lead (Pb) occurs in many forms in the environment; the average

1. J. H. Schulte, "Effects of Mild Carbon Monoxide Intoxication," *Archives of Environmental Health*, 7, no. 5 (November 1963): 524–530. At least one other study has suggested that effects are a linear function of CO concentration beginning at zero. J. M. Heuss, G. J. Nebel, and J. M. Colucci, "National Air Quality Standards for Automotive Pollutants—A Critical Review," *Journal of the Air Pollution Control Association*, 21, no. 9 (September 1971): 535–548.

human intake of lead from food and water in the United States is 0.12 to 0.35 milligrams per day. Lead is gradually excreted from the body; so, at a given rate of intake, an equilibrium body burden will ultimately be reached. Except in cases of occupational exposure, most atmospheric lead comes from automobiles; thus, atmospheric and blood levels of lead are higher in heavily traveled cities than in rural areas.

Much is known about the health effects of large doses of lead, because of occasional acute occupational exposures. For example, the safe limit of ambient lead concentrations set for occupational exposure is over 100 times ambient urban atmospheric concentrations. Stokinger and Coffin conclude that "... there is no present evidence that concentrations in urban atmospheres, where community exposures are the highest, are causing harmful effects."[2] Goldsmith cites no mechanism for injury by ambient concentrations and admits that "... no identifiable cases of toxicity from lead as a community pollutant have been noted," although "... any increase in the body burden of lead is undesirable."[3] Schroeder, however, expresses great concern over increasing lead concentrations in humans and predicts that diseases due to lead toxicity will emerge within a few years.[4]

Unlike carbon monoxide, whose ill effects are felt almost immediately, the toxic damage from excessive exposure to lead accrues over a period of years. Although it is very difficult, at present, to predict exactly what the long-run effects may be, there is reason to fear that lead from automobile exhausts may now be causing or may in the future cause serious health problems. Thus, while the benefits of lead abatement are necessarily more speculative than those of carbon monoxide, they may be equally important in the long run.

HYDROCARBONS In 1967, the Morse Report stated:

No direct health effect is attributable to hydrocarbons at atmospheric concentrations experienced to date. Certain hydrocarbon derivatives emitted

2. Herbert E. Stokinger and David L. Coffin, "Biologic Effects of Air Pollutants," in Arthur C. Stern, ed., *Air Pollution,* 2nd ed. (New York: Academic Press, Inc., 1968), I: 445–546.
3. John R. Goldsmith, "Effects of Air Pollution on Human Health," in Stern, *Air Pollution,* pp. 547, 600, 602.
4. Henry A. Schroeder, "Trace Elements in the Human Environment," report submitted as testimony to the U.S. Senate Subcommittee on Environmental Pollution, Washington, D.C., August 26, 1970.

in automobile exhaust may have carcinogenic effects on lung tissue, but the evidence is inconclusive.[5]

Subsequent research led to a similar conclusion in 1970.[6] Thus the situation of hydrocarbons appears similar to that of lead in that any known direct effects occur at concentrations far above those found in the ambient atmosphere. There is only an unconfirmed possibility that current exposures are harmful, and methodological problems make it unlikely that more will be learned about such low-level effects in the near future.

The primary significance of hydrocarbons for human health lies in their contribution to the production of photochemical smog (PCS), a complex process that is not yet fully understood. Accordingly, to construct a loss function for hydrocarbons, we must first construct one for photochemical smog and then relate hydrocarbon emissions to smog levels. This point will be covered in our subsequent discussion on photochemical smog.

NITROGEN OXIDES Internal combustion engines produce a number of oxides of nitrogen (NO_x), of which the most important are nitrogen dioxide (NO_2), nitric oxide (NO), and nitrogen pentoxide (N_2O_5). These oxides of nitrogen can cause an offensive odor, irritate eyes and nasal passages, cause respiratory diseases, and in large doses can be fatal. There is generally no observable acute tissue response in man or animals until some critical concentration is reached, well above ambient levels.

Until recently, it was thought that ambient concentrations of NO_2 were not sufficient to cause adverse health effects. This is reflected in the Morse Report's statement:

The low concentrations which occur in the community atmosphere have not been identified as damaging to health, but investigations have not been adequate to determine the significance of this pollutant as a public health problem.[7]

---

5. *The Automobile and Air Pollution: A Program for Progress,* Report of the Panel on Electrically Powered Vehicles (Morse Report), (Washington, D.C.: U.S. Department of Commerce, 1967).
6. U.S. Department of Health, Education and Welfare, Public Health Service, *Air Quality Criteria for Hydrocarbons* (Washington, D.C.: U.S. Government Printing Office, 1970), no. AP–64, p. 7–1.
7. Morse Report, p. 14.

A major study in Chattanooga, Tennessee, has shown that in the areas of that city experiencing high levels of NO_2, residents suffer from reduced forced expiratory volume and have a higher rate of respiratory illness than in other areas.[8] While the significance of this study is hotly debated,[9] it does suggest that the direct effects of NO_2 may not be zero at current concentrations.

The second oxide, NO, is at least five times less toxic than NO_2 and produces no irritating reactions. Furthermore, it is very rapidly converted to NO_2, so its concentration in air is well below that of NO_2. For these reasons, NO is not considered to be a health problem. The third oxide, N_2O_5, also occurs at very low concentrations relative to NO_2, and its toxicity seems to be in doubt. There is no evidence of harm at ambient concentrations.[10] The direct effects of NO_x are likely to be more harmful than those of hydrocarbons, but not as harmful as the effects of carbon monoxide.

While its direct effects are not great, like hydrocarbons, NO_x is important because of its contribution to the formation of photochemical smog. At some relative concentrations, a reduction in NO_x will actually increase the PCS produced; while at others, it will lower it somewhat.[11] Small changes in NO_x emissions would therefore produce uncertain results, although large reductions would clearly reduce the PCS formed.

PHOTOCHEMICAL SMOG Photochemical smog is the combination of gases that results from the action of sunlight on motor vehicle exhaust in a warm atmosphere. The primary inputs to the photochemical process are oxides of nitrogen, hydrocarbons, and atmospheric oxygen. The principal products are other hydrocarbons, ozone, nitrogen dioxide, and peroxyacyl nitrates (mainly peroxyacetyl nitrate, or PAN).[12] The effects of PCS include irritation of the eyes and upper respiratory tract and unpleasant odor. At high

8. Carl M. Shy, John P. Creason, Martin E. Pearlman, Kathryn E. McClain, Ferris B. Benson, and Marion M. Young, "The Chattanooga School Children Study: Effects of Community Exposure to Nitrogen Dioxide," *Journal of the Air Pollution Control Association,* 28, no. 8 (August 1970): 539–545.
9. Heuss, Nebel, and Colucci, "National Air Quality Standards."
10. Stokinger and Coffin, "Biologic Effects of Air Pollutants," p. 475.
11. Basil Dimitriades, *On the Function of Hydrocarbon and Nitrogen Oxides in Photochemical Smog Formation,* draft report (Bartlesville, Okla.: U.S. Bureau of Mines, July 13, 1970), p. 22 and Figs. 5, 21.
12. U.S. Department of Health, Education and Welfare, Public Health Service, *Air Quality Criteria for Photochemical Oxidants* (Washington, D.C.: U.S. Government Printing Office, 1970), no. AP–63, p. 2–5.

concentrations, PCS can cause irritation of the skin, respiratory diseases, and great discomfort to the mucous membranes. It appears that these effects rise more than proportionally to the concentration of the PCS products in the atmosphere. PCS is the most likely to be noticed of all the gases resulting from the operation of gasoline-fueled internal combustion engines. It is the primary source of the smog problem that has plagued southern California since the 1940s.

While the process that produces PCS is not completely understood, the form of the primary reactions is known. NO_2 is broken down by the ultraviolet light energy of sunlight into NO and O. The free oxygen atom then combines with an O_2 molecule to form ozone (O_3), one of the primary PCS products. To some extent, the O_3 molecule can then recombine with NO to form oxygen molecules and NO_2. These three reactions will reach an equilibrium concentration of the reagents, in which the O_3 concentration depends in part upon the light intensity and the ratio of NO_2 to NO.

In order for these reactions to produce significant O_3 concentrations, there must also be certain types of hydrocarbons in the air, particularly unsaturated hydrocarbons such as olefins and substituted aromatics. The presence of these hydrocarbons increases the rate of production of O_3 and NO_2, while depleting NO. In addition to producing O_3, the PCS reactions produce peroxyacyl nitrates and other hydrocarbons, mainly aldehydes.

There are many kinds of hydrocarbons, and they vary widely in their ability to participate in the PCS reaction. Various attempts have been made to rank some of them according to their ability.[13] This is particularly important because control programs to reduce hydrocarbons often do not reduce the components in proportion, producing cases where gross hydrocarbons are reduced but photochemical smog productivity is increased. But the measurement of reactive ability is complicated by the fact that the production of PCS depends on the amount of sunlight, which varies seasonally and from city to city.

13. See, for example, Basil Dimitriades, B. H. Eccleston, and R. W. Hurn, "An Evaluation of the Fuel Factor Through Direct Measurement of Photochemical Reactivity of Emissions," *Journal of the Air Pollution Control Association,* 20, no. 3 (March 1970): 150–160; Marvin W. Jackson, "Effects of Some Engine Variables and Control Systems on Composition and Reactivity of Exhaust Hydrocarbons," Society of Automotive Engineers, *Vehicle Emissions, Part II,* 12 (1968): 241–267; and U.S. Department of Health, Education and Welfare, Public Health Service, *Photochemical Oxidants,* p. 2–9.

The problem is to find an index that adequately measures the contribution of hydrocarbons to PCS formation. For our purposes, the index developed by the Bureau of Mines,[14] based on NO_2 and NO_x formation rates, appears to be the most useful available. In evaluating policies that might vary the proportions of the components, we will refer to reactive rather than gross hydrocarbons. In general, some olefins and highly substituted aromatics produce the most oxidant, while acetylene, benzene, and paraffins produce the least.

Production of PCS requires that all three ingredients be present simultaneously. The reaction itself proceeds over a period of several hours, so that it is possible to measure its progress by taking observations at different times of day. A typical daily cycle shows low nighttime levels of all pollutants, with NO, HC, and NO_2 increasing during the early morning rush hour. When the sun is high enough to produce high concentrations of ultraviolet rays, the oxidant concentration will begin to rise, with a simultaneous fall in NO. All inputs and products are likely to peak before noon on a weekday and then decline slowly until nightfall. The relationship is such that it is possible to predict the maximum daily oxidant concentration from the early morning (6 A.M. to 9 A.M.) concentration of hydrocarbons.[15]

During the summer, the intensity and duration of sunlight do not vary enough over the continental United States to affect greatly the production of PCS, although other factors such as cloud cover or other pollutants can have a substantial effect. In winter, however, there are variations of more than a factor of two in the available sunlight for PCS production, according to latitude. In addition, low temperatures inhibit the reaction—for example, a 40-degree drop reduces its speed by a factor of two to four.[16]

Because of the many factors that influence PCS production, the multi-

14. Dimitriades, Eccleston, and Hurn, "Evaluation of the Fuel Factor." An alternative is the General Motors reactivity scale, using a linear summation of the reactivity of each hydrocarbon component. Jackson, "Effects of Some Engine Variables." Both the Bureau of Mines and General Motors have developed eye irritation scales. Some researchers have measured the concentration of oxidants produced from a given amount of the hydrocarbon, while one measured the depth of cracks in rubber produced by the oxidant after a specified exposure. U.S. Department of Health, Education and Welfare, Public Health Service, *Hydrocarbons,* p. 2–9.
15. U.S. Department of Health, Education and Welfare, Public Health Service, *Hydrocarbons,* p. 5–7.
16. U.S. Department of Health, Education and Welfare, Public Health Service, *Photochemical Oxidants,* p. 2–13.

plicity of products that might be measured, and the variety of input concentrations that exists, it is difficult to draw general conclusions about the effect of reductions in inputs on the production of PCS. It appears that, over a wide range, the production of oxidants is proportional to reactive hydrocarbon concentrations, although this is not always true. As we noted earlier, the influence of NO_x is mixed; in some situations a reduction in NO_x will reduce oxidant formation, while in others it may actually increase it. Dimitriades[17] concluded that a reduction in HC from present atmospheric levels would cause a reduction in all components of PCS, except for some nitrates. He also found, not surprisingly, that as HC was reduced, the benefits from NO_x reduction became smaller, since there was less remaining PCS to be abated. In other words, sensitivity to NO_x concentration decreased with HC concentration. Finally, he found that PCS could be formed even with very low HC levels.

The evidence can be summarized as follows. Carbon monoxide is important because of its immediate and direct effects. Hydrocarbons and oxides of nitrogen are important primarily because of their role in the formation of PCS, although they may have some direct effects as well. Lead is suspected as a potential health hazard, at a danger that lies between the possible direct effects of hydrocarbons and oxides or nitrogen.

For all the components of automobile pollution, the length of exposure is as important as the concentration, given the low levels at which exposure occurs. The effects discussed above have been in terms of different concentrations at a given length of exposure. They could also be interpreted in time-dose terms, reflecting the fact that decreased duration may be as beneficial as decreased peak concentration.

One conclusion that can be drawn from the medical evidence is that a given reduction in hours × ppm of exposure is more beneficial at high than at low pollutant levels for all components except, possibly, lead. The test of a program, however, should not be the reduction it will make in maximum concentrations or in *average* concentrations, although these are important; rather, it should be the reduction it will make in time-dose exposure, with greater weight given to reducing high than low levels.

TOTAL HEALTH EFFECT OF AUTOMOTIVE POLLUTANTS

It is useful to identify the maximum economic loss that can be attributed

17. Dimitriades, *Hydrocarbon and Nitrogen Oxides,* p. 21.

Table 3.1. Annual Cost of Pollution-Related Diseases (millions of dollars)

Type of Cost	Cancer of Respiratory	Chronic Bronchitis	Acute Bronchitis	Cold	Pneumonia	Emphysema	Asthma
Death	518	18	6	na	329	62	59
Burial	15	0.7	0.2	na	13	2	2
Treatment	35	89	na	200	73	na	138
Absentee	112	52	na	131	75	na	60
Total	680	159.7	6.2	331	490	64	259

Source: Ronald G. Ridker, *Economic Costs of Air Pollution* (New York: Frederick A. Praeger, Inc., 1967), p. 54.

to automotive air pollutants, as a guide to the maximum amount that should be spent on abatement. While this approach is not as desirable as the specification of a loss function, it can offer some guidance for policy when it is combined with the preceding information on the shape of the loss function.

One relevant estimate has been made by Ridker.[18] He defines the loss caused by a disease as the lost wages due to premature death, the premature cost of burial, the cost of treatment, and the wages lost because of absenteeism. Medical evidence suggests that seven diseases may be caused by air pollution: cancer of the respiratory system, chronic bronchitis, acute bronchitis, the common cold, pneumonia, emphysema, and asthma. Using a 5 percent discount rate, Ridker has determined annual resource costs associated with these diseases (see Table 3.1).

Because of the blanks in the table, these costs are considered to be an underestimate, as is the total—$1,989 million per year in 1958. To arrive at the maximum percentage of this total cost that may be attributed to air pollution, Ridker notes the difference between urban and rural rates of death for these diseases. If the rural rates had been used, 20 percent fewer deaths would have been recorded. While this does not allow for occupational, socioeconomic, or other differences between urban and rural populations, it does provide a reasonable estimate. His total cost attributable to air pollution is thus $400 million per year. Inflating this figure by 20 percent for population growth and 30 percent for inflation since 1958, we de-

18. Ronald G. Ridker, *Economic Costs of Air Pollution* (New York: Frederick A. Praeger, Inc., 1967).

rive a total of $3.1 billion for 1970, $620 million of which is attributable to air pollution. This too is probably underestimated, since health-care costs have risen far more than the consumer price index.[19]

More recent studies have used multiple regression analysis to test the relationship between air pollution and diseases. Hodgson[20] has analyzed the incidence of a number of diseases in relation to sulfur dioxide and particulates in New York between November 1962 and May 1965. He finds a significant correlation between particulates and heart and respiratory disease, such that an increase in the daily average particulate concentration by one coefficient of haze (COH) unit [21] for one month raises the average daily death rate for that month by 13.35 deaths per day. Since he does not attach a dollar value to this result, however, and because it is difficult to relate COH to automobile pollution, this study sheds little light on the costs of auto emissions.[22]

Lave and Seskin[23] have conducted one recent study of pollution and health. They review a number of articles that suggest quantitative relationships between specific pollutants and diseases, and perform their own econometric analysis of the data from other studies. They conclude that a 50 percent reduction in air pollution would reduce bronchitis morbidity and mortality by 25 to 50 percent, lung cancer mortality by 25 percent, respiratory disease morbidity and mortality by 25 percent, and all mortality from cancer by 15 percent. By applying to the figures a set of costs for treatment and lost earnings from illness and premature death, they estimate that the 50 percent reduction in urban air pollution would be worth at least $2.08 billion per year. This is over three times the amount derived

19. Feldstein shows that two measures of hospital costs approximately doubled from 1960 to 1968 (see Martin S. Feldstein, "The Rising Cost of Hospital Care," Discussion Paper no. 129, Harvard Institute of Economic Research, August 1970). This would justify more than doubling the hospital care portion of the total loss, making the overall inflation factor about 50 percent and the 1970 air pollution cost $720 million.

20. Thomas A. Hodgson, Jr., "Short-Term Effects of Air Pollution on Mortality in New York City," *Environmental Science and Technology*, 4, no. 7 (July 1970): 589–597.

21. Coefficient of haze: a measure of obscurity of the air.

22. One study employed a detailed biological model to link air pollution to health, but did not reach quantitative conclusions that can be used here. Richard J. Hickey, D. E. Boyce, E. B. Harner, and R. C. Clelland, "Ecological Statistical Studies on Environmental Pollution and Chronic Disease in Metropolitan Areas of the United States," Discussion Paper no. 35, Regional Science Research Institute, University of Pennsylvania, 1970.

23. Lester B. Lave and Eugene P. Seskin, "Air Pollution and Human Health," *Science*, 169, no. 3947 (August 21, 1970): 723–733.

from Ridker's figures, because of Ridker's lower cost estimates per death or per day of treatment for the diseases and his attribution of only 20 percent of the incidence of disease to air pollution. If Ridker's figure, adjusted to 1970, is doubled to reflect the 25 to 50 percent incidence that Lave and Seskin attribute to air pollution, the cost of air pollution is raised to $1.24 billion per year, and then to $1.5 billion with the health care cost adjustment.

On the basis of these two studies, then, the value of health losses caused by the excess of urban over rural air pollution can be calculated at between $1.5 and $2 billion.

Both of these studies seem sound in the methodology of their approach to the problem. Unfortunately, they use particulates and sulfur dioxide as the pollution indicators, and motor vehicles do not cause a significant portion of either of these. If motor vehicle pollutants are perfectly correlated with particulates and sulfur dioxide, then they may be at least partial causes of the observed diseases, but there is no way of separating their effect. If they are independent, then the existing studies are not relevant to a determination of the health effects of automotive pollutants. If, as is most likely, there is a partial correlation between the pollutants studied and motor vehicle pollutants, then further work is necessary to determine the effects attributable to each.

The Lave and Seskin study may be used to determine the maximum amount that could be attributed to automobile pollution, given their methodology. First, we will assume that they have correctly identified all diseases that might be caused by air pollution, whether from other sources or from automobile emissions. The maximum harm attributable to pollution would be 100 percent of the relevant diseases. If the costs cited in the study are used, the total loss from pollution-related diseases would be about $14.4 billion per year. This, then, is an absolute maximum on the health benefit from known diseases that could be achieved by reduction of automobile pollution. The most likely value would be only a small fraction of this, since much disease is not related to air pollution at all and some is related to nonautomobile pollution. It seems quite improbable that the amount attributable to automobile pollution reduction could be larger than $3.4 billion (25 percent of the maximum), and more likely that it is close to $1.3 billion (10 percent). This suggests that the health benefits from complete abatement of automobile pollution would be valued at a few billion dollars per year.

OTHER EFFECTS

The evidence relating to other effects of automobile air pollutants is similar to that on health effects in that there is a large amount of detail, but little upon which to base estimates of aggregate damage or loss. For example, consider aesthetic effects, which, in the case of automobile pollution, consist primarily of the irritation and brown haze of photochemical smog. It has been estimated[24] that an oxidant component of 0.15 ppm causes slight eye irritation. Pasadena and Los Angeles exceed this value on 41 and 30 percent of all days, respectively. San Diego exceeds 0.15 ppm on 5.6 percent of all days, and Denver, on 4.9 percent of all; all other cities are at 2.5 percent or less.[25] Multiplying the population of each of these four cities by the percentage (using 10 million as the population of the entire Los Angeles basin and 33 percent as the incidence there) gives the number of person-days per year of exposure to above 0.15 ppm: 1.24 billion.

The real problem lies in converting these exposure figures into dollar values. Studies do not exist that establish a solid value for the concern of individuals over exposure to irritating levels of photochemical smog. To choose an arbitrary figure, if half the exposed population were willing to pay $1 per day to avoid the irritation alone, aside from health effects, the total value of a reduction below the critical level for these four cities would be $620 million per year. This pollution level could be reached by a 50 percent reduction in emissions, since only 5 percent of the hours in a year have concentrations greater than 0.18 ppm in Pasadena and the yearly average concentration there is only 0.042 ppm.[26] Thus if these figures are used, the aesthetic loss appears to be about one-half the value of the health loss. Such comparisons are not very useful, however, since the value of a person-day of discomfort is without basis in empirical research.

Photochemical smog has long been known to be injurious to plants as well as to man. In grape leaves, ozone causes a stippling effect, cell walls collapse, and a brown pigment is produced. Tobacco leaves also exhibit

24. U.S. Department of Health, Education and Welfare, Public Health Service, *Photochemical Oxidants,* Chap. 9.
25. Ibid., pp. 3–2 and 9–19.
26. Ibid., p. 3–3.

changes in color and damage to the cell structure in the presence of ozone. Numerous crops and even some pine trees have been found to be susceptible to ozone damage. PAN can cause damage to leaves, principally around the stomata, which will lead to discoloration and some separation of the leaf tissue.

In general, the heaviest concentrations of automotive pollutants are in urban areas, so that plant damage tends to affect ornamental rather than agricultural plants. One exception to this rule is Southern California, where the pollution from the Los Angeles basin covers substantial acreage of agricultural land.

The aggregate effect of continued exposure to high PCS levels in this area has been a reduction in the growth and yield of important cash crops. Crop damage in Southern California has been estimated at $44 million per year.[27] Citrus crops are in decline, with early loss of leaves, smaller fruit, and poor growth. Vegetable production also is affected. In many cases, yields may be reduced without any of the visible damage effects referred to above. It is clear, however, that plants vary in their susceptibility to PCS damage; some species may suffer seriously, while others are essentially unaffected. The most sensitive species are often used as indicators to show when pollution has reached the injury threshold level.[28]

On the basis of the California statistics, total crop damage from automobile pollution is much less than health damage, since few parts of the country have so severe a problem with photochemical smog; and rarely does that problem extend beyond the urban area to valuable agricultural land. For example, it has been estimated that air pollution damage to all kinds of vegetation in Pennsylvania in 1969 was $3.5 million and that oxidants, produced largely through automobile pollution, accounted for 83 percent of this total, or $2.9 million.[29]

Photochemical smog, particularly ozone, has been clearly shown to be a significant cause of damage to certain kinds of materials. Some organic

27. Quoted in a news broadcast from Dr. Nichols, Director of Plant Pathology, California Department of Agriculture, February 6, 1971.
28. U.S. Department of Health, Education and Welfare, Public Health Service, *Hydrocarbons,* Chap. 6.
29. T. Craig Weidensaul and N. L. Lacasse, "Results of the Statewide Survey of Air Pollution Damage to Vegetation," *Journal of the Air Pollution Control Association,* Paper 70–108, 63rd Annual Meeting, St. Louis, June 1970.

compounds, especially rubbers and textiles, are particularly sensitive to deterioration by ozone, which attacks long polymer chains, reducing tensile strength and sometimes increasing brittleness. In fact, automobile tires, which account for a large proportion of domestic rubber production, are made of compounds that are especially susceptible to ozone damage. The damage occurs when the rubber is stretched; this causes a protective film on the surface to break and allows the ozone to penetrate to the subsurface, which, in turn, will crack allowing further penetration. Since tires are constantly flexed, they have no protection. The amount of damage depends upon the type of rubber, the time-dose of ozone, the stress, and the temperature. Thus the less PCS, the less damage; and the lower the temperature, the less damage. It would appear that in cooler areas of the country there would be less damage, even at constant PCS concentrations. Also, it has been suggested that the rubber composition of tires can be modified to resist ozone, at a cost of about $0.50 per tire.[30]

Many kinds of fabrics have been found to weaken and fade in the presence of ozone. The mechanism of loss of strength seems to require that the fabric be wet when exposed, reducing the number of situations in which damage is likely. Fading also may depend on humidity, although this point is not so certain. It appears that fabrics and dyes can be selected that will resist ozone damage, but some cost is likely to be associated with this selection. Present data are not sufficient to calculate costs for the damage or its avoidance.

Given these additional effects of automobile pollution, benefits of abatement should include reduced maintenance of vehicles, reduced materials damage, and reduced laundry and other cleaning bills. We do not yet have the data required to evaluate this function. For the present, it must be included informally, since the benefits from improved health understate the total benefits from automobile pollution reduction.

Spatial and Temporal Considerations

So far, the benefits from pollution abatement have been considered either on a microscopic physiological level or on a very aggregated level—viewing national benefits from a uniform reduction in pollution everywhere.

30. U.S. Department of Health, Education and Welfare, Public Health Service, *Photochemical Oxidants,* p. 7–2.

In practice, however, it is important to consider two more dimensions to the problem: geographic variations in the loss function related to population density or degree of pollution, and changes in the loss function over time. It does not appear that the world atmospheric burden of automotive pollutants is increasing at a rate sufficient to cause concern. The problems associated with automobile pollution are in the area near the source, where the effluent density is high, before the pollutants have settled out, dispersed, or been converted into harmless gases. Thus we must determine how the benefit from a given unit of abatement varies according to the location of the emitting vehicle.

It appears that the primary harm caused by automotive pollutants is injury to human health and the aesthetic offense to human senses. The other damage, except for crop loss, is done to personal property, which is located wherever people live. This suggests that the harm from a given unit of pollution should be proportional to the number of people who are exposed and whose property is exposed to it. If air quality were uniformly poor throughout the entire country but population were still concentrated in the cities, then a given unit of pollution would affect more people in cities because more would breathe and rebreathe it, and more would smell it, see the haze, and be irritated by photochemical smog. In short, even with a uniform pollution density, the loss per unit of emission would increase with increasing population density.

Furthermore, air pollution is not uniformly distributed. The most heavily polluted areas are in and around the great urban population centers. In recalling the discussion of medical effects, one can conclude that the harm from automobile pollutants rises more than proportionally to their concentration in the air. If the loss function were linear with pollution density, then the local air quality would be irrelevant to the determination of benefits from a given unit of abatement. With this increasing marginal loss, the benefits from a unit of abatement are greater in dirty areas than in clean areas. Thus, even if the population were spread evenly across the country, the benefits from abatement in dirty areas would be greater than the benefits from the same quantity of abatement in cleaner areas.[31]

31. Other economists have shown that a city with more pollution and more population should incur greater marginal abatement costs, but its optimal air quality can be either

While this relationship cannot be quantified, it is clear for two reasons that the benefits from a given unit of reduction in emissions are greater in urban than in other areas: because of the greater population density; and because of the higher concentration of pollution. This suggests that, in addition to any policies aimed at reducing the emission rates of new cars sold throughout the country, local or state pollution control programs should be initiated that are responsive to local and state population densities and pollution levels. Formulation of such policies can be aided by a description of the variation of pollution densities over time and space.

In urban areas, near streets and expressways, the automobile pollution level is almost directly proportional to traffic volume. It peaks during the morning and evening rush hour, declines by midday, and is much lower at night; concentrations of lead, for example, are 40 percent higher in the daytime than at night.[32] In residential areas, it may be nearly constant at perhaps one-quarter the average urban value.[33] Concentrations are highest in vehicles moving in traffic, somewhat lower at the curb of the same street (perhaps half as much), and lower still at points even farther from the highway source.[34] The lead found in dirt and soil also decreases rapidly in concentration as distance from the highway increases.[35] Thus, it appears that the concentration of some automotive pollutants can vary by a factor of two to four within a few hundred feet, depending on the position relative to the highway source. Since the ambient level of carbon monoxide in the air evidently is not increasing over time, it appears that there must be some natural disposal process.[36]

cleaner or dirtier than the other city. Sam Peltzman and T. N. Tideman, "Local versus National Pollution Control: Note," *American Economic Review*, 62, no. 5 (December 1972): 959–963.

32. J. M. Colucci, C. R. Begeman, and K. Kumler, "Lead Concentrations in Detroit, New York and Los Angeles Air," *Journal of the Air Pollution Control Association*, 19, no. 4 (April 1969): 255.

33. George D. Clayton, Warren A. Cook, and W. G. Fredrick, "A Study of the Relationship of Street Carbon Monoxide Concentrations to Traffic Accidents," *Industrial Hygiene Journal*, 21 (1960): 46–54.

34. Robert M. Brice and Joseph F. Roessler, "The Exposure to Carbon Monoxide of Occupants of Vehicles Moving in Heavy Traffic," *Journal of the Air Pollution Control Association*, 16 (November 1966): 597–600.

35. Colucci, Begeman, and Kumler, "Lead Concentrations."

36. Louis S. Jaffe, "Ambient Carbon Monoxide and Its Fate in the Atmosphere," *Journal of the Air Pollution Control Association*, 18, no. 8 (August 1968): 534–540.

These findings have several implications for automobile pollution control strategies. First, the large variations over very small distances suggest that some problems may be reasonably attacked on a very local basis. Blocking off one street to traffic may reduce the pollution level on that street by 50 percent, even though the average air quality for the city will be essentially unaffected. Furthermore, if the world atmospheric burden of automotive pollutants is not accumulating over time—and apparently it is not in the case of carbon monoxide—then we are dealing with a flow and not a stock. A reduction in the rate of emissions will cause a reduction in ambient pollution, not a reduction in its rate of increase. Thus, control strategies can be based upon an understanding of the current problem in polluted areas. We can look for benefits in the short-range terms that have been considered here, instead of speculating on the effect of continuous accumulation.

We must not overlook the fact, however, that the population continues to grow, the automobile fleet continues to grow, and the number of automobile miles grows most rapidly of all (at a rate of about 5 percent per year in the United States). If the rate of emissions per vehicle-mile remains constant, even though there is no long-term accumulation of automotive pollutants, air quality will still deteriorate as the number of vehicle-miles increases. Furthermore, because the urban population is growing, the harm done by even a constant level of pollution will increase over time; if pollution increases, its harm will increase even more rapidly.

Clearly, even if we can determine the optimal rate of emissions for a given point in time, that rate will not be optimal at some point in the future. As the marginal loss rises, so also does the optimal marginal cost of abatement, and the optimal emission standard will constantly increase in stringency. Conversely, if a strict new standard causes air quality to improve rapidly, in a few years the optimal standard may not be so strict. Thus, when standards are set for new cars, it should be remembered that it will be five years before half of all cars on the road meet that standard, and both air quality and population density change in the meantime. New car standards should reflect, not current conditions, but an average of the conditions expected over the life of the car, which is about ten years. At present, with new cars producing less pollution than older cars, the average urban concentration of automotive pollutants is steady or falling. If the current standards are optimal for the present, it will be difficult to justify stricter standards until air quality begins to deteriorate again or until technical progress lowers the marginal cost of abatement.

Implications for Abatement Objectives

Selection of an optimal pollution control strategy requires specification of the benefits of pollution abatement as a function of all relevant parameters. The evidence suggests that benefits are a function of the density of each of the four pollutants and of population density or of total affected population. It is assumed that the nonhealth effects are proportional to population, since they arise mostly from impact on population-related activities, except for crop losses. Thus, for a given airshed, the loss from pollution could be expressed as

$$L = f(X_i, Z) \quad \text{for } i = 1 \text{ to } 4, \tag{3.1}$$

where L is the total loss, X_i is the rate of emission of a pollutant (i), and Z is a population variable. Since it appears that for some pollutants the marginal loss is an increasing function of density, the functional form might properly include an exponential treatment of X_i.

Unfortunately, the evidence does not give an accurate total loss figure or sufficient information that would permit the selection of a functional form and estimation of numerous parameters. While this precludes derivation of a true benefit function, it may still be possible to determine the relative importance of the four pollutants and thereby derive an abatement index that can be used for comparison of policies varying the proportions of the four pollutants. Indeed, unless such an index can be constructed, analysis and comparison of policies are impossible.

The creation of a pollution index requires several assumptions. Starting with Equation 3.1, we assume that over the relevant range the effects of each pollutant are linear with concentration (this does not contradict our earlier finding of nonlinearity, provided that only a small range is considered). Second, we assume that these effects are independent of each other. Third, we assume that they are additive. Then a pollution index could be written as

$$P = Z + \sum_{i=1}^{4} a_i X_i,$$ (3.2)

where a_i is the relative weight assigned to each pollutant. If policies can be segregated and compared for identical situations, then the population term Z is unnecessary, and the index is

$$P = \sum_{i=1}^{4} a_i X_i.$$ (3.3)

This index is useful for comparing changes in emissions that are a small percentage of total emissions and that occur in areas of uniform population density. When these assumptions are not valid for a given situation, adjustments must be made in the use of the index.

The only remaining problem is to determine the weight of each pollutant. The health data do not present a clear case for a unique set of values. It appears, however, that in areas where photochemical smog is not a problem, carbon monoxide is probably a more serious health problem at ambient concentrations than the other three. Where PCS is a problem, CO, HC, and NO_x may be equally important and lead somewhat less important. For PCS-plagued areas, if one took the position that the first three pollutants were equally important and lead was half as important in the concentrations found in the urban air, then allowed for the fact that automobiles contribute about half of the first three and all of lead in the air, the appropriate weights would be the reciprocal of the emissions rate from an uncontrolled automobile. In other words, a 10 percent reduction in HC would be valued as equal to a 10 percent reduction in CO, NO_x, or lead.

This assumption leads to a single figure of merit consisting of the algebraic sum of all increases and decreases from the pre-device emission rates per mile of the four pollutants expressed as percentages of the base levels, divided by four. This figure will be referred to as the "Net Percent" measure of abatement. Thus we have

$$\triangle \text{Net Percent} = \left(\sum \frac{\triangle \text{CO}}{\text{Base CO}} + \frac{\triangle \text{HC}}{\text{Base HC}} \right.$$
$$\left. + \frac{\triangle \text{NO}_x}{\text{Base NO}_x} + \frac{\triangle \text{Pb}}{\text{Base Pb}} \right) \div 4. \tag{3.4}$$

By this measure, strategies that reduce one pollutant but raise another by the same percentage of its base level make no improvement. The base level used for reference is the uncontrolled automobile.

In a sense, this Net Percent measure of abatement seems arbitrary, since automobiles may emit quite different proportions of pollutants, under the same conditions of loss; thus the result is a different weighting of changes in individual pollutants. In another sense, it is not arbitrary, however; if one cannot distinguish among the four at a given set of relative concentrations, and the probable harm of each is proportional to concentrations, then the concentrations at which probable harm is equal are the present ones. Given current data, it seems unlikely that any other figure of merit could be proved to be better.

Net Percent is a measure of pollution per mile and is therefore useful for comparing strategies which reduce pollution per mile. In addition, we need some means of measuring the abatement caused by reductions in total motoring. For this purpose, the same assumptions are used as for Net Percent, and the total emissions from an uncontrolled automobile when driven 10,000 miles (one year) are defined as one Pollution Unit (PU), consisting of equal contributions of the four primary pollutants. When motoring is reduced, the pollution reduction is the emission rate of the vehicle divided by base emission rates, times the number of miles reduced over 10,000; thus we have

$$\triangle \text{Pollution Units} = \frac{\text{Emission rate}}{\text{Base emissions}} \times \frac{\triangle \text{ mileage}}{10,000}. \tag{3.5}$$

Assuming that vehicles are driven 10,000 miles per year, Pollution Units are numerically equal to Net Percent divided by 100.

A third figure of merit can be derived, again by using the same assumptions but with the added proviso that risk should be avoided in the selection of an abatement policy. A change is not an improvement unless it is certain to leave us better off than before. This is analogous to Pareto-Superior states, in which nobody must be worse off and some must be better off. Given the uncertainty about the effect of each pollutant, improvement must be measured by the reduction in the least-abated pollutant. If we totally eliminated three pollutants but left the fourth unchanged, the Net Percent would record a large improvement, and the expected value of the total loss would be greatly reduced; but if most of the loss in fact resulted from the fourth pollutant, conditions might well remain the same as they were before abatement. The figure of merit for this objective function is the percentage reduction, from base levels, of the least-abated pollutant. It will be referred to, hereafter, as Minmax Percent. Using Minmax Percent as the measure of abatement will not achieve the maximum expected benefit per dollar invested, but it will produce the maximum certain benefit per dollar invested. It is, as we have suggested, a very conservative or risk-averse measure:[37]

$$\frac{\text{Minmax}}{\text{Percent}} = \text{MIN} \left(\frac{\triangle \text{CO}}{\text{Base CO}}, \frac{\triangle \text{HC}}{\text{Base HC}}, \frac{\triangle \text{NO}_x}{\text{Base NO}_x}, \frac{\triangle \text{Pb}}{\text{Base Pb}} \right). \qquad (3.6)$$

A fourth measure assumes that only those pollutants affected by the strategy in question appear in the loss function and that the others are irrelevant. This attributes to those who set the standard or require the device an accurate understanding of the true loss functions. It then gives the cost of the sum of the percentage reduction for those pollutants that are reduced. This measure, which we refer to as Percent Reduced, is useful primarily for comparing strategies that affect only the same pollutant (unless one is prepared to believe in omniscient decision makers!). Where more than one pollutant is reduced, the cost is allocated among them, depending upon the circumstances of the particular case:[38]

37. Each term in the equation is included only if it is negative.
38. Each term in the equation is included only if it is negative.

$$\text{Percent} \atop \text{Reduced} = \sum \left(\frac{\triangle CO}{\text{Base CO}} + \frac{\triangle HC}{\text{Base HC}} \right.$$

$$\left. + \frac{\triangle NO_x}{\text{Base NO}_x} + \frac{\triangle Pb}{\text{Base Pb}} \right). \tag{3.7}$$

Finally, reductions in lead emissions are frequently achieved by means quite different from those used to reduce the other pollutants. Therefore two more indices are constructed, similar to Net Percent and Minmax Percent, but excluding lead completely:[39]

$$\text{Unleaded} \atop \text{Net} \atop \text{Percent} = \left(\sum \frac{\triangle CO}{\text{Base CO}} + \frac{\triangle HC}{\text{Base HC}} + \frac{\triangle NO_x}{\text{Base NO}_x} \right) \div 3; \tag{3.8}$$

$$\text{Unleaded} \atop \text{Minmax} \atop \text{Percent} = \text{MIN} \left(\frac{\triangle CO}{\text{Base CO}}, \frac{\triangle HC}{\text{Base HC}}, \frac{\triangle NO_x}{\text{Base NO}_x} \right). \tag{3.9}$$

These measures are useful for comparing strategies that do not control lead or for comparing other strategies when lead is considered unimportant. Minmax Percent, for example, is useless for comparing strategies that leave one pollutant unchanged, if that pollutant is included in the analysis.

There are many cases in pollution control and elsewhere when formulation of sensible policies founders on the problem of constructing a meaningful measure of output or improvement. For air pollution control, some have proposed using a health effects index, in which each pollutant is weighted by its relative impact on human health. If it were possible to construct a loss function that was analyzed by pollutant and include dollar values, this could be used as the objective function for abatement. Since this cannot be done, we have attempted to devise a working alternative by assigning equal weights to equal percentage reductions for three or four pollutants. Some alternatives also could be constructed based on some other set of assumptions about the relative importance of the four components.

39. Other indices could be derived, of course, giving lead a weight somewhere between the zero of Equations 3.5 and 3.6 and that of Net Percent.

Our method is clearly superior to using total grams or tons of pollution, since there is no reason to believe that the loss caused by a pollutant is proportional to its molecular weight.

Conclusions

The data describing the impact of automobile air pollution on man and his environment are incomplete and not sufficient to permit us to specify a precise loss function for this pollution. But we have uncovered some useful facts about it. It appears that carbon monoxide may affect cognitive and psychomotor performance of motorists at some of the higher ambient concentrations now existing and that the shape of the dose-response curve is exponential. Lead, hydrocarbons, and oxides of nitrogen have not been proved to have adverse short-term health effects at present concentrations; but in the medical community, it is feared that they may cause long-term injury, resulting in increases in disease rates and reductions in life expectancy. The photochemical products of automotive emissions are known to be unpleasant, irritating, and unhealthy at levels frequently encountered in some areas. And some damage to materials and plant life is known to occur as a result of automobile emissions. The total cost of health effects may be on the order of a few billion dollars per year. Still, these estimates are so uncertain that we must rely more on cost-effectiveness analysis than cost-benefit analysis.

It is concluded here that, with the uncertainty about individual effects, one reasonable assumption is that each is equally harmful, with the possible exception of lead, at present concentrations. This leads to the construction of two basic objective functions or indices of pollution abatement. The first records the average reduction from base levels of the four pollutants and is termed Net Percent. The second, Minmax Percent, is more risk-averse and records only the percentage abatement from base levels of the least-abated pollutant. Thus a strategy that reduces some pollutants but not others will have some effect on the first index but none of the second. Other indices also could be supported, given the current state of our knowledge about the effects of the four components.

Because of variations in population density and pollution density, together with the increasing marginal loss with density of pollution, the benefits from a given unit of abatement are greater in densely populated than in sparsely populated areas and greater in dirty than in clean areas. Thus, the benefits per unit of abatement in large urban areas may be much greater than elsewhere. This suggests that whatever national programs of pollution abatement are followed, there is also a need for state and local programs to respond to the particular needs of individual areas. For the same reasons, as population and motoring grow over time, the optimal degree of abatement will increase, even without technical progress in abatement technology.

Costs of Pollution Abatement: Methodology

4

The earliest estimates of the costs of automobile pollution control were based primarily on the increase in new car prices attributable to pollution control devices. Since then, it has become apparent that pollution controls can have a significant impact on vehicle maintenance costs and on fuel consumption. Some controls affect the performance of the vehicle in important ways and should not be ignored. Finally, the total cost of controls is rising to the level where it might have a significant impact on consumers, affecting the choice of vehicle, new car sales, or the amount of motoring. This chapter will specify what assumptions were used to compute abatement costs, including assumptions about the behavior of consumers and firms and about the elements that enter into the cost calculation.

Behavioral Models

CONSUMERS

In the past, automobile pollution control programs generally have been analyzed on a ceteris paribus basis, on the assumption that motorists will not alter their behavior as a result of the controls. But the technological and regulatory alternatives required to make further improvements may be much more costly than previous measures, and the effectiveness of several of them will rely specifically upon motorist response. An evaluation of these strategies must therefore be grounded upon some understanding of how households make decisions about their motoring and other transportation activities, and how they will react to the forces that new controls will create.

The household decision-making process can be divided into two parts: vehicle purchase and vehicle operation. These decisions are closely interrelated. For example, the decision to make a particular trip by automobile is dependent upon a prior decision to buy (or rent) an automobile. Still, the decisions are made at different points of time and rely on quite different information.

In general, we are making the standard assumptions about consumers: that they are utility maximizers; that they have perfect information and foresight; and so on. These assumptions imply that, when a consumer purchases a vehicle, he knows what the stream of future operating costs will be for that vehicle (and for any others he has considered) and that he discounts the operating cost stream along with his expected stream of satisfaction with the vehicle. Clearly, these are strong assumptions. People may be accurate at ranking cars in order of their expected operating cost; but it is unlikely that they can attach dollar figures to fixed and variable costs, and even less likely that they will properly discount them. In fact, many households probably cannot state very accurately the costs of owning their present car, much less divide those costs into fixed and variable components. Therefore, in cases where these assumptions are important for evaluating a strategy or comparing alternative strategies, their validity will be questioned more closely and the consequences explored.

Strong as these assumptions are, they need not impair the analysis. Even if motorists do not know the dollar amounts of ownership and operating costs, they probably know what factors enter into the computation and how. Their response to changes in cost structure would thus be in the direction predicted by theories based upon these assumptions. Only the magnitude of the response would be unpredictable, and even this might not be hard to estimate for aggregate populations. Thus, if each purchase decision did not accurately reflect the present value of the operating cost stream, purchase decisions probably would respond quite predictably to changes in that stream. It follows that if we want to value changes in operating cost or to predict responses to them, we must do so from evidence based upon operating cost data rather than upon variations in purchase price.

Finally, consumer tastes are taken as unchanging. This is a common assumption in microeconomic analysis, but it is particularly important to note its impact here. It means that the demand for motoring and for vehicle attributes such as power will be assumed not to shift in response to changes in environmental quality or perceptions of that quality. It means that people will not be assumed to be willing to forgo trips just to reduce urban congestion and pollution. In the short run, this assumption of constant tastes is probably quite accurate because of the enormous effort ordinarily required to alter them. In the long run, however, it may be seriously questioned. Life-styles can change over time, and it appears that the current generation of youth may have life-styles and values significantly different from those of their parents. While it is too early to make reliable predictions, this may have some impact on the demand for motoring, or on the type of vehicle that will be preferred a decade from now. It should be recognized that the conclusions reached here may be more pessimistic, and the abatement costs higher, than will be experienced in a world growing more concerned about preserving the existing environment.

FIRMS

It is assumed that the major firms involved in automobile pollution matters, the four domestic automobile manufacturers,[1] are long-run profit maximizers, and that in the short run they maximize profits subject to some broad constraints, such as avoiding the risk of antitrust suits. They are progressive and zealous in lowering production costs, but only moderately innovative; much of the new technology comes from outside sources. They are little motivated by factors that do not appear in the marketplace—that is, by automotive features for which there is little customer demand or profit potential. This claim seems to be borne out by their record of performance in the areas of automobile safety and pollution, both of which have been characterized by slow technical and marketing progress. Product development seems to follow oligopolistic patterns, with none of the big three[2] entering a new submarket, such as compact cars, until there is enough demand to support all three.[3]

The primary form of competition is product differentiation, with emphasis on styling. Pricing follows a traditional pattern of oligopolistic price leadership, with General Motors taking the lead. The price level, however, seems to be flexible, with annual changes and some upward or downward modifications during the year. While there are some differences in profit rate among different models produced by a single manufacturer, they tend to correlate with price; the compacts and subcompacts bring lower rates than full-sized cars. It can be assumed that prices generally are proportional to manufacturing costs. There are substantial differences in profit margin among the four manufacturers, who rank, from highest to lowest, as follows: General Motors, Ford, Chrysler, and American Motors.[4]

1. General Motors, Ford, Chrysler, and American Motors.
2. General Motors, Ford, and Chrysler.
3. Lawrence J. White, *The Automobile Industry Since 1945* (Cambridge, Mass.: Harvard University Press, 1971).
4. For exhaustive analysis of the automobile industry, including specific treatment of pricing, safety, and pollution, see White, *Automobile Industry Since 1945*.

Cost Elements

The costs of the pollution control devices considered here fall into three categories: capital costs, reflecting an increase in the cost of the new vehicle; maintenance costs, reflecting a change in the cost of maintaining the engine or the device itself; and operating costs, which vary with changes in fuel consumption rates or prices. Current prices are used, so if fuel prices increase exogenously, these costs will increase. The cost considered is social cost, which equals the costs to motorists less changes in tax payments, adding any major governmental administrative or enforcement costs not charged directly to users.

In general, vehicle performance is unchanged, so that the costs apply to the production of vehicles with the same safety, comfort, size, and so on, but altered pollution output. Where an important vehicle parameter, such as horsepower, is changed, a value is imputed to that change.

CAPITAL COSTS

Many new car strategies alter the design of the vehicle in a way that increases its production cost. It is assumed in the preceding section that the automobile industry is characterized by price leadership, but that there is sufficient competition that prices bear a proportional relationship to manufacturing costs. It is further assumed that most firms experience constant or slightly declining costs. These assumptions imply that an increase in manufacturing costs will be passed directly on to consumers by the industry as a whole, although different cost increments among firms may be masked.

In addition to direct costs in the form of higher new car prices, pollution controls can cause costs of a less visible kind by changing the performance of the vehicle. In recent years, many models have displayed annoying characteristics such as hard starting, stalling, and hesitation in acceleration. While reduced drivability (to use Detroit terminology) does not introduce direct monetary costs, there is a loss of satisfaction on the part of the motorist which theoretically may be given a dollar value; in fact, this becomes a real cost in cases where the motorist pays to have pollution control devices disconnected in the hope of improving his vehicle's performance.

In principle, all changes in vehicle performance resulting from pollution controls should be valued and included as a cost of the pollution control program. Problems of drivability, however, do not lend themselves to precise measurement and evaluation; and while they may be important to motorists, they cannot as a practical matter be evaluated here. A change that *can* be valued is the loss in power that has accompanied recent pollution controls, and that will undoubtedly increase as stricter controls are imposed. Horsepower is clearly an important vehicle attribute; manufacturers have advertised it for decades, and motorists have happily paid premiums for automobiles with extra amounts of it. The usual economic assumption that goods are homogeneous is discarded here; the automobile is treated as a multiattribute durable good, where the demand for a particular attribute can be identified and a value for that attribute estimated. We will use least-squares regression analysis to relate new car prices to a set of parameters that are thought to represent, directly or indirectly, the characteristics that determine the value of automobiles. This incorporates the method of hedonic quality measurement developed by Griliches[5] and others.[6]

If a control device lowers the rated horsepower, it is necessary to value that loss. One might assume that people would always buy the same amount of the parameter and thus note the added cost of producing the same horsepower as before with the addition of the control device. This suggests that the price of that attribute will rise. But with most commodities, if the price rises, the amount purchased falls. Here we should use the demand curve for the attribute to predict the amount that will be purchased at the new price and then value the increase in production cost plus lost consumer surplus.

In Appendix B, it is shown that for recent makes and models, and under certain conditions and assumptions, power is priced at about $1.69 per horsepower. A price elasticity of demand for power also was determined to be about −0.17. The satisfactory fitting of linear price equations validates the hypothesis that when a control system changes technology to shift the price of power upward, it does so in a linear fashion, so that average and marginal costs remain equal and constant. The social cost of such a shift in the supply curve is the increase in price for the new quantity purchased plus the lost consumer surplus (see Figure 4.1).

5. Zvi Griliches, "Hedonic Price Indexes for Automobiles: An Econometric Analysis of Quality Changes," in U.S. Cong., Joint Economic Committee, *Government Price Statistics,* Hearings . . . January 24, 1961 (Washington, D.C.: U.S. Government Printing Office, 1961), pp. 173–196; and "Hedonic Price Indexes Revisited: Note on the State of the Art," *Proceedings,* Business and Economics Statistics Section, American Statistical Association, 1967, pp. 324–331.
6. See, for example, Jack E. Triplett, "Automobiles and Hedonic Quality Measurement," *Journal of Political Economy,* 77, no. 5 (1969): 408–417.

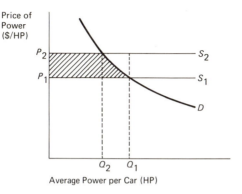

Figure 4.1. Cost of Changing Price of Power

Algebraically, this can be written as

$$C = (P_2 - P_1)Q_2 + \tfrac{1}{2}(P_2 - P_1)(Q_1 - Q_2) \qquad (4.1)$$

and

$$C = \tfrac{1}{2}(P_2 - P_1)(Q_2 + Q_1). \qquad (4.2)$$

The first term in Equation 4.1 is the shaded rectangular area in Figure 4.1, representing the increase in production cost for the attribute in question given the new supply curve S_2, which cuts quantity to Q_2. The second term is the triangular shaded area under the demand curve, which equals the lost consumer surplus arising from the shift, if one assumes a linear approximation to the demand curve. The horizontal supply curve reflects the assumption of constant marginal cost of power production, which is therefore equal to average cost.

For calculation of the cost of a particular pollution control device, the available data are the original price and quantity of power purchased (P_1 and Q_1) and the increase in production cost, expressed as a percentage of the original cost. If the percentage increase in production cost is L, and the price elasticity of demand for power is E, then P_2 will be $P_1(1 + L)$. And Q_2 will be the original amount of power times one plus the price elasticity of demand times the price change, or $Q_2 = Q_1(1 + E \times L)$. Substituting these values in equation 4.2 gives

$$C = \tfrac{1}{2}P_1Q_1L(2 + E \times L). \tag{4.3}$$

Because of the long literature debating the desirability of using consumer surplus in project evaluation,[7] it is interesting to see how important that surplus is in this calculation of the cost of power loss. Using the elasticity of demand of -0.17, an initial price of $1.69 per horsepower, and assuming a 5 percent increase in the cost of producing power, we get a total cost of $19.47. Of this amount, consumer surplus accounts for $\tfrac{1}{2} \times 2 \times 0.0875 =$ $0.0875. With such a highly inelastic demand, the consumer surplus loss is negligible compared to the increase in production cost. The primary importance of the price elasticity of demand for power, then, is to estimate how much power will be purchased after an increase in its price.

Capital costs, including the change in new car cost (aside from power) plus the cost of power price changes, are spread over the life of the vehicle instead of being converted to present values, because of the difficulty of placing a dollar value on benefits and the conceptual problems in discounting physical benefits. The capital cost is treated as a lump sum investment to be repaid over the life of the vehicle in equal annual installments. With annualization, annual or per-mile costs are compared with annual or per-mile pollution reductions.

A 6 percent discount rate is used for the annualization.[8] Over a ten-year

7. Pungent criticism is offered by Paul A. Samuelson, *Foundations of Economic Analysis* (New York: Atheneum, 1967), pp. 195 ff.; and I. M. D. Little, *A Critique of Welfare Economics* (Oxford: Oxford University Press, 1960), Chap. X.

8. Since the costs considered here fall on the owner or operator of the vehicle, they are in the private sector, and the appropriate discount rate is thus some average private rate. A minimum figure would be the real rate of return on private investment, which cannot exceed productivity increases, or 3 percent per year. A maximum would be the private borrowing rate, which ranges from 6 to 12 percent for most types of consumer credit. Since over one-third of all new cars are bought on time (see Automobile Manufacturers Association, *1970 Automobile Facts and Figures,* Detroit, Michigan, 1970), and of the remainder many are probably financed by some source other than the vendor, the borrowing rate is weighted more heavily to arrive at the 6 percent figure. All rates are in real (no inflation) terms. For a general discussion of time discounting, discount rates, and opportunity costs, primarily for public sector investment, see Martin S. Feldstein, "Net Social Benefit Calculations and the Public Investment Decision," *Oxford Economic Papers,* 16 (March 1964): 114–131; "The Social Time Preference Discount Rate in Cost-

period (the average life of an American automobile),[9] 6 percent gives an annual cost equal to 0.136 of the original capital cost. It is then possible to convert annual costs to mileage costs, approximating the average U.S. figure of 9,600 miles per year with 10,000 miles per year.[10] Because mileage accumulates more rapidly in early years than in later ones, greater benefits and depreciation costs should be found in the early years. But since both benefits and costs are shifted, a linear approximation should not seriously affect the results. When a capital improvement is made on a used car, the cost of that improvement is amortized over the remaining life of the vehicle.

Annual capital costs (AK) are thus

$$AK = (K + C) \times 0.136, \tag{4.4}$$

where K is the cost of control devices and C the cost of lost power from Equation 4.3.

Obtaining accurate cost estimates for the various devices or design changes is a problem in this analysis. Even existing devices are hard to evaluate. Since they are not sold as an extra cost option, the portion of the price attributed to them by the manufacturer is arbitrary. Moreover, there are clear incentives to overstate this cost, to mask a general price increase. In a few cases, independent price estimates are available. But where they are not, probable prices can be estimated on the basis of a description of the device and the cost of replacement parts. In the case of proposed devices, many are now operating as prototypes, and current sales and installation costs are sometimes available. It should be noted, however, that these figures are only indicators of the cost of installing particular devices as standard equipment on all new cars.

MAINTENANCE COSTS

Generally, little information is available on maintenance costs for vehicles or on the impact of various pollution control devices on these costs. We will

Benefit Analysis," *Economic Journal*, 74 (June 1964): 360–379; and "Financing in the Evaluation of Public Expenditure," Discussion Paper no. 132, Harvard Institute of Economic Research, August 1970.

9. J. A. Fay and M. Scott Mingledorff, "Dynamics of Automobile Population and Usage," Working Paper no. 2, Legislative Research Conference, Columbia University, April 18, 1972.

10. Automobile Manufacturers Association, *1970 Automobile Facts and Figures*, p. 56.

assume, then, that maintenance costs are unchanged unless the pollution control device specifically requires some periodic adjustment, cleaning, or replacement. This will tend to understate the cost of some devices or overstate their effectiveness in cases where lack of the specified maintenance will cause deterioration in performance of the vehicle. Even some design changes that do not add any equipment to the vehicle may increase maintenance costs. For example, progressive leaning of the air-fuel ratio during the last few years requires that this ratio and all engine tune-up parameters be adjusted more precisely, since the leaner mixture is closer to a mixture on which the engine will not run; and more precise adjustment may mean more frequent and more expensive tune-ups. Maintenance costs, as potential hidden costs of pollution abatement, are an important area for further engineering research. When they are quantified, these costs are expressed on a per mile basis. Thus, if PM is the cost of a maintenance operation and FM is its frequency during the life of the vehicle, then annual maintenance cost (AM) is

$$AM = \frac{PM}{FM}.\tag{4.5}$$

OPERATING COSTS

Operating costs, as distinguished from maintenance costs, are the cost of consumables, principally gasoline. Since many abatement devices can have substantial effects on gas mileage, this is a potentially large cost element. Fuel consumption and price changes are evaluated by deviation from the 1970 average of 15 miles per gallon (MPG) for the U.S. automobile fleet[11] and the 1970 average gasoline price of $0.35 per gallon, including tax. Of this $0.35, about $0.11 is federal and state gas tax,[12] so that the untaxed cost of a change in fuel consumption is only $0.24 per gallon. If gasoline prices increase greatly in the future, then the cost of policies that increase fuel use is underestimated.

If previous fuel consumption is MG miles per gallon, and it increases by

11. See Appendix A.
12. U.S. Department of Transportation, Bureau of Public Roads, *Highway Statistics 1968* (Washington, D.C.: U.S. Government Printing Office, 1970).

G percent, with user price at UPG and a tax of TG dollars per gallon, the annual cost is

$$\text{AFU} = \frac{10^4}{\text{MG}} \left(\frac{G}{1 - G} \right) \text{UPG} \qquad \text{(User)}, \qquad (4.6)$$

$$\text{AFS} = \frac{10^4}{\text{MG}} \left(\frac{G}{1 - G} \right) (\text{UPG} - \text{TG}) \qquad \text{(Resource)}. \qquad (4.7)$$

If the cost of fuel changes by R dollars per gallon, assuming a constant tax rate, then annual cost is

$$\text{AFU} = \text{AFS} = \frac{10^4}{\text{MG}} \times R. \qquad (4.8)$$

The total cost of both changes together would then be

$$\text{AFU} = \frac{10^4}{\text{MG}} \left[\left(\frac{G}{1 - G} \right) \times (\text{UPG} + R) + R \right], \qquad (4.9)$$

$$\text{AFS} = \frac{10^4}{\text{MG}} \left[\left(\frac{G}{1 - G} \right) \times (\text{UPG} + R - \text{TG}) + R \right]. \qquad (4.10)$$

We will refer to the cost with tax as the user cost, since it is the one to which users respond. Strategies will be evaluated, however, on the basis of the pretax price, or the resource cost. The large income transfer from motorists to state and federal governments that results from these taxes means that any change in gas mileage has significant transfer effects, as well as resource effects. We will assume that on balance, the marginal gasoline tax dollar has the same value in the hands of motorists as in the hands of the government which collects the tax. To do anything else would require vast analyses of the efficacy of government spending generally, and of highway investment specifically, for the many cases where gasoline taxes are earmarked for highway purposes. Note, however, that since motoring expenditures rise much less than proportionally to income, a gasoline tax—like any sales tax—will be regressive in its incidence. We consider a regressive redistribution of income to be undesirable, and we will therefore downgrade any plan that results in significant increases in gasoline tax receipts.

It could be argued that, aside from the gasoline taxes, gasoline is underpriced, because of the large tax concessions given to petroleum companies through the depletion allowance, expensing of intangibles, and other means, or because limited reserves are not reflected in prices. Against this there is the argument that foreign oil is less expensive than domestic, generally; that the oil industry is an oligopoly and may charge more than competitive prices; and that state and local taxes on mineral extraction more than make up for the income tax break.[13] On balance, the pretax retail price must be accepted as a fair representation of the resource cost of the fuel.

ADMINISTRATIVE COSTS

Any program to control automobile pollution will involve some administrative costs for testing vehicles, enforcement, and perhaps gathering data. For example, it can cost almost as much to measure precisely the emissions from a vehicle as it does to install many of the current control devices. Whether a program requires testing of new cars at the factory, periodic emission measurements on older cars, or highway surveillance with testing of apparent violators of some standard, if the costs of the testing are significant, they should be included in the cost of the program. These costs may accrue directly to motorists—for example, new cars may be tested by the manufacturer and the cost of the test included in the price, or a periodic inspection may be required that must be paid for by the motorist. Or they may accrue first to the regulatory agency—as in the case of government testing of new cars, and random testing of vehicles on the highway. In a sophisticated model, different shadow prices could be assigned to expenses, depending upon who had to shoulder their burden. But following the assumption here that a marginal dollar in the hands of any government has the same value as in the hands of an average motorist, we will not be concerned with such differentiation.

The total annual cost of any pollution control device can now be ex-

13. For an analysis of petroleum pricing, see Edward W. Erickson and Leonard Waverman, eds., *The Energy Question, Multinational Economic Policy* (Toronto: Toronto University Press, 1974). Since October 1973, foreign oil has become more expensive than domestic.

pressed as the sum of capital, maintenance, operating, and administrative costs, either with or without tax. Total annual user cost is

$$\text{TACU} = \text{AK} + \text{AM} + \text{AFU}$$

$$= [K + \tfrac{1}{2}P_1Q_1L(2 + L \times E)] \times 0.136 + \frac{\text{PM}}{\text{FM}}$$

$$+ \frac{10^4}{\text{MG}} \left[\left(\frac{G}{1-G} \right) \times (\text{UPG} + R) + R \right]. \tag{4.11}$$

Total annual resource cost is

$$\text{TACS} = \text{AK} + \text{AM} + \text{AFS}$$

$$= [K + \tfrac{1}{2}P_1Q_1L(2 + L \times E)] \times 0.136 + \frac{\text{PM}}{\text{FM}}$$

$$+ \frac{10^4}{\text{MG}} \left[\left(\frac{G}{1-G} \right) \times (\text{UPG} + R - \text{TG}) + R \right]. \tag{4.12}$$

In the relation, above, it is assumed that a vehicle is driven 10,000 miles per year over a ten-year life.

Fuel Consumption, Fuel Composition, and Pollution

5

It is sometimes suggested that automotive emissions should be controlled by regulation directed at some parameter other than the composition of exhaust gases as they leave the vehicle, on the assumption that this other parameter will directly or indirectly reduce the harmful content of those gases. For example, in 1970 the California Legislature considered a bill to tax new cars by engine size, with penalties for large engines.[1] This bill was based on the assumption that larger engines meant higher pollution rates. This assumption, in turn, derives from a two-stage reasoning process: larger engines mean greater fuel consumption, and more fuel consumption means more pollution.

Here we will first examine the relationship between gross vehicle parameters, including engine size, vehicle weight, and fuel consumption, to quantify this relationship as precisely as possible. We will then assess the hypothesis that fuel consumption and pollution emissions may be proportionally or otherwise related. Finally, we will explore the effect on emission rates of variation in gasoline composition and substitution of other fuels, which is another subject of indirect regulation that has received substantial attention. The analysis of fuel consumption determinants is useful not only for evaluating pollution policy but also for evaluating ways to save fuel in the event of an energy crisis, which would mean substantial price increases.

Engine Size, Vehicle Weight, and Fuel Consumption

Appendix A reports on a study of the relationship between some gross vehicle parameters, including power and weight, and the rate of automobile fuel consumption in North American driving conditions. Multivariate regression analysis is used to relate gas consumption to a few significant explanatory variables in several data sets representing groups of automobiles whose characteristics are known and whose gas mileage has been measured. The results show that, for a fixed set of driving conditions, engine size (measured either by power or by displacement) and vehicle weight can explain over 80 percent of all variations in gas mileage. The results can be used to determine the difference in fuel use and operating cost, which will depend on selection among available engine and body sizes.

---

1. See R. d'Arge, T. Clark, and O. Bubik, "Automotive Exhaust Emission Taxes: Methodology and Some Preliminary Tests," Project Clean Air, Research Report S-12, University of California at Riverside, September 1970.

Alternative equations present the relationships that were discovered. A linear equation, accurate in the middle of the range of vehicle specifications, representing medium-sized cars, is in the form

$$MPG = 28.3 - 0.00178WT - 0.217HP, \tag{5.1}$$

where MPG is miles driven per gallon of gas, WT is gross vehicle weight, and HP is maximum-rated engine horsepower. A logarithmic form was also estimated and is accurate over a wider range of vehicle specifications. The underlying algebraic equation is

$$MPG = \frac{Const \times (1 + 0.0625CR)}{WT^{0.273} \times HP^{0.370}}, \tag{5.2}$$

where CR is the compression ratio of the engine. Other equations, some of which use engine displacement instead of horsepower, are presented in Appendix A, Table A.1.

In Equation 5.1 it is implied that a weight increase of one pound will reduce gas mileage by 0.00178 miles per gallon, while a power increase of one horsepower will decrease gas mileage by 0.217 miles per gallon. In a typical vehicle weighing 3,700 pounds with a 250-horsepower engine, a 10 percent reduction in gas mileage is effected either by raising the weight by 800 pounds (21 percent) or by raising power by 65 horsepower (25 percent). Thus, reducing the average vehicle in the fleet from the 1968 size (which represents a full-sized Ford, Chevrolet, or Plymouth with a medium-sized V-8 engine) to a small compact (by eliminating 1,000 pounds and 100 horsepower) would increase gasoline mileage by 4 miles per gallon, or about 25 percent.

Equations 5.1 and 5.2 are based on data sets from vehicles manufactured between 1966 and 1968. Since 1970, design changes resulting largely from increasingly strict pollution controls have reduced the amount of horsepower attainable from a given displacement and have increased fuel consumption for a vehicle of a given horsepower or displacement. These changes undoubtedly will continue until emission rates stabilize and engine technology settles down again. To some extent, these changes will be captured in the equations that include compression ratio explicitly, since compression ratio reductions account for a large part of the change. Beyond this, however, fuel economy will be worse than these equations suggest, although the reduction may be a simple proportion of the estimate.

One implication of these findings is the variation in total life-time fuel consumption that results from changes in these parameters. The reduction of power by 100 horsepower increases average gas mileage from 15 to 17 MPG. If a car lasts ten years and is driven 10,000 miles per year (estimates are 9,500), at 15 MPG it will consume 667 gallons of gas per year, while at 17 MPG it will consume only 588 gallons, or 79 gallons per year less. Using the 1970 gasoline price of 35 cents per gallon, this is an individual saving of $27.65 per year on gasoline alone. Discounted at 6 percent for ten years, it would total a present value of about $200. This is about 6 percent of the price of a new automobile, and more than the added new car price for 100 extra horsepower. If the price of gas rose to 50 cents per gallon, the annual saving would be $39.50, or $286 over the life of the car.

Fuel Consumption and Pollution Output

The use of these data to predict vehicle emissions from fuel consumption assumes that there is a strong connection between the rate of fuel use and the rate of pollution output among vehicles, if abatement technology is held constant. An investigation of that relationship is now in order, to see whether the initial assumption is correct and how significant the relationship is. The task is complicated by the fact that there is no easy way to determine whether technology is held constant. The most sophisticated controls in use on 1971 cars included over a dozen modifications and additions to the engine, and a particular engine may have had any combination or subset of these modifications. Indeed, different engines made in a single year by one manufacturer may incorporate different technology.

The problem is that since 1968 all domestic (large-engine) automobiles and since 1970 all automobiles have had to meet the same emission standard measured in grams of pollutant per mile driven. Manufacturers do whatever is necessary to make each engine-transmission-body combination meet this standard, and no more. Naturally dirty combinations, such as manual transmission cars, require more sophisticated treatment than naturally clean combinations. Many manual transmission cars have been equipped with the rather expensive air injection pump to add air to the exhaust manifold, while few nonmanual transmission models require this system. Thus, if the manufacturers adequately perform their job of just meeting the standard with each automobile, tests will show no systematic variation in emissions with gross vehicle parameters. In short, while technology in any given year is not constant, emissions may be.

Logic and the results of the gas mileage regressions can provide some guidance, however, as to what emissions would be *if* technology were uniform. If we hold vehicle size (weight) constant and vary the engine size, we find a variation in fuel consumption of about two miles per gallon for every 100 horsepower. The larger the engine, the more easily it can carry the vehicle through any particular driving cycle. It has been said that small engines have trouble meeting air pollution standards because, in working as hard as they must, they cannot run efficiently. In other words, a small engine may be tuned for maximum power per displacement to give adequate performance, while a larger engine, with excess power capacity, may be tuned to minimize emissions, which is definitely not power maximizing. Also, decreasing the surface-to-volume ratio of the combustion chamber decreases hydrocarbon pollution. For a given number of cylinders, as the displacement increases, the surface-to-volume ratio decreases; this tends to make large engines cleaner. There are thus some reasons for large engines to be cleaner and some for small engines to be cleaner; it is not possible to specify a priori the relationship between engine size and rate of emissions.

Another explanation for the lack of correlation between engine size and emissions may be that technology is proportional to engine size—that is, more effort is made to clean up large engines than small ones. This may result automatically from efforts to equalize emissions if uncontrolled emissions are greater from large than from small engines. It may result also from the manufacturers' desire, not to spend a constant amount per vehicle for abatement (which could be the constant technology assumption), but to spend less on small cars. This desire would be a natural result of the higher price elasticity of demand for small than for large cars. This is only speculation, however, and a survey of manufacturers would be necessary to determine whether or not the assumption of constant technology is valid for vehicles of various engine sizes that must meet the same emission standard.

While engine size may not be reliably related to emissions, the compression ratio certainly is. If other design parameters are held constant, raising the compression ratio increases peak combustion temperatures and thus raises the production of oxides of nitrogen. There is also a small tendency toward a reduction in HC emissions as the compression ratio is raised; and the composition of these emissions changes somewhat because of different burning conditions for the fuel. In addition, the compression ratio is a prime determinant of the octane that the fuel must maintain to avoid knocking; raising fuel octane requires either more lead or more reactive hydrocarbons, under most refinery processes. Thus, if compression ratio alone were varied, use of an appropriate fuel would result in variations in both NO_x and reactive HC emissions, both rising with rising compression.

Let us now hold engine size constant and vary vehicle weight. As weight increases, the engine must work harder to achieve the same performance. This will raise peak combustion temperatures, increase nitrogen oxide emissions, and thus require heavier use of power jets in the carburetor, increasing hydrocarbons and carbon monoxide. Furthermore, fuel consumption will rise by two miles per gallon per 1,000 pounds of weight. Since there is no evidence that an engine operating at low power levels is particularly dirtier in pollution per gallon of fuel burned than an engine that operates at high power levels, we would expect that pollution output would increase more or less in proportion to the increase in fuel consumption. Thus, while engine size may have an indeterminate effect on pollution output, vehicle weight will probably raise the level of pollution, for a fixed engine size, about as much as it raises fuel consumption.

One potential source of information about the relationship between gross vehicle parameters and pollution output is studies of vehicles built before exhaust controls were installed. While many measurements have been made on such automobiles, there have been no analyses to date that relate the pollution output to gross vehicle parameters. If emission measurements on such cars have been made, and if at the same time the major gross vehicle parameters were recorded,[2] it would be most interesting to do a complete statistical analysis and try to extract the underlying relationships.

A study similar to that proposed has been conducted at the University

2. Frequently such tests record the make and model of the automobile, but no engine data. Since most automobiles offer a wide variety of engines, it is almost hopeless to specify later what engine was in a particular vehicle.

of California.[3] Unfortunately, the data were from tests on cars built from 1966 to 1969, which in California all included exhaust controls. No attempt was made to define or classify the technology of control used on each car. Furthermore, the explanatory variables included the number of cylinders, engine displacement, gas mileage, rated engine horsepower, and compression ratio. The correlation coefficients for these five variables tended to be about 0.75, with displacement and horsepower at 0.96, and displacement and gas mileage at 0.90. Since all five highly collinear variables were left in the major equations tested, it is not surprising that they were rarely significant and that the signs of the estimated coefficients were not always easy to explain. Vehicle weight was not included. The results are cited for the proposition that engine size is not strongly related to pollution emission rates. It is true that the study does not establish the converse; but it is suggested here that a definitive answer will not be reached until a study includes vehicle weight as an explanatory variable, includes only one or two measures of engine size, and uses a vehicle fleet in which abatement technology is, in fact, constant.

A third source of information about the relationship between gross vehicle parameters and pollution output comes from casual observation of the clean air car race of August 1970 run between Cambridge, Massachusetts, and Pasadena, California. Entrants generally had some choice of vehicle and engine, which they modified in a variety of ways. While technology was, by definition, not constant among the entrants, it is significant that the teams did not seem to have a preference for small engines as a base. Many of the cars were powered by V-8s where a six-cylinder engine would have been available. In at least one case, a 307-cubic-inch eight-cylinder engine was removed and a 350-cubic-inch eight-cylinder engine installed in its place.[4] Several entrants cited the advantages of having a big slow-turning engine rather than a small hardworking one. On the other hand, no car used the largest available engines; even those of high displacement were equipped with carburetors and adjustments that produced only modest horsepower. This suggests that there is no relationship between displace-

3. R. d'Arge, T. Clark, and O. Bubik, "Automotive Exhaust Emission Taxes: Methodology and Some Preliminary Tests," Project Clean Air, Research Report S-12, University of California at Riverside, September 1, 1970.
4. This was the Worcester Polytech's Propane Gasser, which was the cleanest car in the race. Another winner used a 302-cubic-inch Ford V-8.

ment and pollution output in the small to medium-size range, while a positive relationship may exist at high power and displacement levels.

Fuel Composition

The amounts of all four pollutants and the content of the hydrocarbon emissions can be affected by altering the fuel that is burned in the engine. Possible changes include varying the volatility of the gasoline, its hydrocarbon composition, or its lead content, or replacing gasoline entirely with gaseous fuels such as propane (LPG) or natural gas (CNG). The chemistry of fuels and their combustion could fill an entire book; all that will be attempted here is an outline of the primary impact of these changes on evaporative and exhaust emissions, and the significance of these facts for pollution control policy.

Fuel volatility is the tendency of the fuel to evaporate and is measured by the Reid Vapor Pressure (RVP), which increases as volatility increases. Reducing the RVP of a fuel from 10 to 5.8 greatly reduces evaporative losses and, at the same time, increases exhaust hydrocarbons, with a total effect of reducing total gross hydrocarbons by up to 26 percent.[5] If, instead of gross hydrocarbons, the measure is a reactivity index indicating the tendency of the hydrocarbons to form photochemical smog, the reduction is only 19 percent. A reduction in RVP from 9 to 6 would cost from $0.013 to $0.023 per gallon of gasoline in extra refining costs.[6] It would also tend to cause starting problems in some areas, particularly in cold weather. The impact on emissions of other pollutants is not large.

These results are for vehicles not equipped with evaporative controls, which have been mandatory since 1971 on all U.S. cars and which, by themselves, reduce evaporative emissions by 80 percent or more. On a vehicle with such evaporative controls, the emission reduction arising from fuel volatility changes would be so small that it would be almost canceled by the increase in exhaust emissions. Reducing volatility thus appears to be a productive way to reduce hydrocarbon emissions only where evaporative

5. Basil Dimitriades, B. H. Eccleston, and R. W. Hurn, "An Evaluation of the Fuel Factor Through Direct Measurement of Photochemical Reactivity of Emissions," *Journal of the Air Pollution Control Association*, 20, no. 3 (March 1970):150–160.
6. U.S. Department of Health, Education and Welfare, Public Health Service, *Control Techniques for Carbon Monoxide, Nitrogen Oxide, and Hydrocarbon Emissions from Mobile Sources* (Washington, D.C.: U.S. Government Printing Office, 1970), p. 6–3.

controls have not been installed in the vehicle, or until such time as most vehicles have them.[7]

Instead of, or in addition to, changing the RVP of a fuel, it is possible to vary the type of hydrocarbons, affecting both evaporation and exhaust emissions. Elimination of the light olefins through C_5 or C_7 has little effect on total gross hydrocarbon emissions but can reduce hydrocarbons measured by the Bureau of Mines reactivity index by 23 and 27 percent, respectively.[8] Surprisingly, this reduction comes almost entirely from reductions in evaporative losses, rather than exhaust reductions; so that, again, it will be much less in vehicles equipped with evaporative control devices. Thus it appears that reductions in volatility or light olefin content below current Los Angeles standards offer only moderate hope for reducing photochemical smog in the short run, and much less in the long run, if evaporative control devices continue to be required.[9]

The removal of tetraethyl lead from gasoline, whether to reduce lead emissions or to avoid poisoning the catalysts that may be used in future abatement systems, tends to raise hydrocarbon emissions. When a premium gasoline was modified to provide the same octane without lead as it had previously had with lead, gross total hydrocarbon emissions increased slightly, but hydrocarbons measured by the Bureau of Mines reactivity index increased by 17 to 21 percent. The reactivity of exhaust emissions increased by 20 to 29 percent, so that evaporative controls would not negate or even reduce this undesirable effect.[10] This increase in reactivity results from the increase in aromatics needed to provide the specified octane level.

When lead was removed from regular fuel, two results occurred, depending on what other changes in composition were made. Going from leaded regular to a low-olefin unleaded gasoline increased the reactivity by only 4 to 7 percent. Going from leaded regular to a high-olefin unleaded fuel produced a 24 to 25 percent increase in total reactive hydrocarbons. Therefore it appears that removal of lead from gasoline can result in great in-

7. James W. Daily, "Los Angeles Gasoline Modification: Its Potential As An Air Pollution Control Measure," *Journal of the Air Pollution Control Association,* 21, no. 2 (February 1971): 79.
8. Dimitriades, Eccleston, and Hurn, "Evaluation of the Fuel Factor," p. 157.
9. John L. Laity and James B. Maynard, "The Reactivities of Gasoline Vapors in Photochemical Smog," *Journal of the Air Pollution Control Association,* 22, no. 2 (February 1972): 102.
10. Dimitriades, Eccleston, and Hurn, "Evaluation of the Fuel Factor," p. 158.-

creases in photochemical smog potential, unless some action is taken to ensure that engines can use low-octane fuel and that fuel has a low-olefin composition. This would account for the fact that, as Detroit manufacturers have introduced engines designed for low-lead and unleaded fuels, they have also reduced compression ratios, so that these engines can run on regular gasoline. The cost of producing low-lead and unleaded gasolines is a complicated function of refinery design and product mix that cannot be summarized adequately here; it is the subject of a separate study.[11]

A substantial reduction in all emissions can be obtained by converting a gasoline engine to run on gaseous fuels such as propane (LPG) and natural gas (CNG). A test of six 1967 to 1969 vehicles converted to run on either gasoline or natural gas showed carbon monoxide reduced by 86 percent, hydrocarbons down by 35 percent, reactive hydrocarbons down 90 percent, and oxides of nitrogen down 64 percent (see Table 5.1). The

Table 5.1. Emissions with Natural Gas Fuel [a] (concentration in exhaust)

Engine Displacement	CO		HC				NO_x	
	Concentration		Concentration		Reactivity		Concentration	
	Gasoline (%)	CNG (%)	Gasoline (ppm)	CNG (ppm)	Gasoline (RU)	CNG (RU)	Gasoline (ppm)	CNG (ppm)
134	1.28	0.19	343	188	1243	153	1270	343
199	2.78	0.11	363	90	2020	201	848	359
250	1.06	0.16	333	146	1838	183	601	152
289	1.17 [b]	0.13	282 [b]	108	1370	86	1529	340
302	0.71	0.21	307	158	2398	238	546	147
440	1.96	0.14	168	298	1643	223	481	352
Average Percent Reduced (%)	86.4		34.9		89.7		63.9	

Source: R. W. McJones and R. J. Corbeil, "Natural Gas Fueled Engines Have Lower Exhaust Emissions," *Society of Automotive Engineers Journal*, 78, no. 6 (June 1970): 32.
a. Data are continuous except reactivity units (RU) which are from flame ionization detector tests on bag samples. All data are averages of readings at conversion and 4,000 miles later.
b. Hot start.

11. The Department of Commerce undertook a comprehensive study of this problem through the Technical Advisory Board Panel on Automotive Fuels and Pollution. A draft report "The Implications of Lead Removal from Automotive Fuel" was issued in June 1970.

large difference between gross hydrocarbons and reactive hydrocarbons occurs because natural gas consists largely of methane, a nonreactive hydrocarbon. The reduction in oxides of nitrogen is possible because natural gas permits the use of a very lean mixture without poor drivability or surging.

Conversion of conventional vehicles to CNG operation requires installation of two or more large high-pressure tanks, usually in the trunk of the car. Because two tanks (which occupy a large amount of trunk space), provide a range of only about 100 miles, this conversion would not be convenient for most drivers. Furthermore, unless stations were built to fill tanks with CNG, the motorist would be severely limited in the places where he could use the car. It appears, then, that this conversion would be most appropriate for urban fleet vehicles that return to a central garage to refuel and do not need a long cruising range. Conversion can be made at a cost of about $350 per vehicle, leaving the gas tank in place for use if the CNG should be exhausted out of range of a refueling station.[12]

From this discussion, it is clear that fuel composition can have an impact on emissions that is, to some extent, independent of vehicle design. These alternatives can be included in the evaluation of abatement choices by determining the effect of each on total emissions (in terms of one of the indices developed in Chapter 3) and calculating the cost of implementation. The resulting cost effectiveness can then be compared with other strategies to determine which of them are worth pursuing.[13]

Conclusions

For vehicles produced under a given technology,[14] the gas mileage for a given driving cycle can be predicted accurately from two vehicle parameters: gross weight and maximum rated engine horsepower. The most reliable equation is

$$MPG = \frac{Const \times (1 + 0.0625CR)}{WT^{0.273} \times HP^{0.370}}.$$

12. Costs and performance of this system are explored more fully in Chapter 7.
13. See Chapter 7 for a comparison of the most promising choices, low-lead fuel and natural gas.
14. That is, in one country and over a short time period.

An easier form to work with, which is accurate in the vicinity of the average vehicle's specifications, is

$$MPG = 28.3 - 0.00178WT - 0.217HP.$$

Either equation will produce an R-squared of over 0.80 for relatively new cars under typical U.S. driving conditions.

In other words, for a vehicle of average size, an increase in weight or in engine size will reduce gas mileage; conversely, a decrease in either of these parameters will increase mileage. This gives a basis for computing the change in operating cost caused by an abatement strategy which, directly or indirectly, changes gross parameters.

A secondary conclusion is that not only is gas mileage greatly affected by driving style, driving cycle, and temperature, but so is sensitivity to the major determinants of gas mileage: weight and power. Thus, when an equation for determining gas mileage is estimated, it is important to specify as precisely as possible the mix of vehicles tested, the driving cycle used, driver habits, and other relevant test conditions.

The impact of variations in fuel consumption caused by gross vehicle parameters is less certain. It has not been established that emissions increase with engine displacement, at constant abatement technology; in fact, for small to medium-sized engines, the converse may be true. It seems quite likely that for a given engine size, reducing vehicle weight will decrease emissions; so direct regulation of weight would be more productive than regulation of engine size. Increasing the compression ratio increases nitrogen oxides, reduces hydrocarbons, but increases hydrocarbon reactivity. Even if emissions were directly proportional to fuel consumption, a significant reduction in the size of the average U.S. car from a full-sized, low-priced car to a compact would reduce fuel consumption, and thereby pollution, by only 25 percent. In fact, the abatement from such a change undoubtedly would be even less. Since variations in engine design and abatement technology affect pollution more than do gross parameters, it does not appear reasonable to regulate the gross parameters strictly for the purposes of pollution control.

Finally, changes in fuel alone can significantly affect emissions. Changing gasoline composition can lower emitted hydrocarbons, particularly reactive hydrocarbons, although the largest impact is on evaporative emissions that can be dealt with by vehicle design changes. Switching to natural gas can significantly reduce all emissions, and may be desirable for urban fleet vehicles.

The Demand for Motoring 6

In recent years, it has frequently been suggested that, in addition to the adoption of other pollution control strategies, the total amount of motoring should be reduced. In the first place, past and current control programs have not improved air quality as rapidly as some had hoped; there is also a general feeling that our dependence on the automobile in urban areas has other undesirable consequences aside from its contribution to air pollution.

Any scheme intended to reduce motoring must recognize the two basic determinants of automobile use: the decision to buy an automobile, and the decision to use it at a particular time. A variety of policy instruments may be employed to affect these decisions and thereby to influence the demand for motoring. In the short run, for example, there may be imposed a tax on new car sales, an annual tax on automobile ownership, a tax or toll on actual use, a tax on one factor (such as gasoline), or a prohibition or restriction on use at certain times in certain places; or a substitute means of transportation may be subsidized or improved. In the long run, investment in motoring facilities (such as highways) may be reduced, land-use patterns may be altered to reduce trips for which the automobile is dominant, or more facilities may be built for substitute means of transportation.

In order to analyze the impact of using any or all of these instruments, one would have to construct an elaborate model of automobile purchase and use that includes all fixed and variable costs, descriptions of urban land-use patterns, the supply of motoring facilities, and the supply of other transport services. Unfortunately, it is not possible to build a reliable model on the basis of the available data. A time series would have to omit World War II because of rationing and the cessation of automobile production; it could not go back too far because accurate cost data do not exist, and technology has changed so much that cost proxies like the prices of a few factors would be quite misleading. Postwar time series on new car sales, prices, miles driven, highway facilities, and transit use reveal primarily smooth trends with few wiggles to display particular responses.[1]

Accordingly, we will have to use a different approach. First, we will review existing studies of demand in three categories: new car sales; intercity

1. White, for example, tried to predict postwar automobile sales and found such a high degree of multicollinearity among the explanatory variables that regression analysis was impossible. See Lawrence J. White, "The American Automobile Industries in the Postwar Period" (Ph.D. Thesis, Department of Economics, Harvard University, 1969), p. 135.

motoring; and urban motoring. Then we will construct a model to explain the distribution of motoring costs and gross consumer reactions to changes in them. The results that can be obtained from an analysis of this type admittedly are rather crude, but they will provide an adequate basis for evaluation of various demand instruments. Although it may not be possible to quantify precisely the impact of each instrument, we will be able to judge its feasibility and relative effectiveness in substantially reducing automobile use on a local, metropolitan, or national scale.

Demand for Automobile Purchases

One source of information about the demand for motoring is data on automobile purchases. While purchase price is only part of the total cost of motoring, the response of sales to changes in price should give some information on the impact of expensive pollution control devices or tax strategies on new car sales and thus, ultimately, on the size of the automotive fleet.

Suits[2] studied new car sales between 1929 and 1956, using real disposable income, the real retail price of new automobiles and a measure of credit conditions as explanatory variables, and units sold as the dependent variable. He used annual data, omitting the wartime period of 1942–1945. Suits found an income elasticity of demand of 4.2; the price elasticity was positive (0.4) and became negative only when the price variable included a measure of credit availability. This short-run price elasticity of demand for new automobiles seems unreasonable; and, since first differences are used throughout, there is no long-run price elasticity. While the income elasticity is a short-run figure, it is large enough to suggest that long-run income elasticities are probably substantially greater than one.

Chow[3] studied the purchase of new and used automobiles from 1921 to 1953. He used as a dependent variable the sale of "new car equivalents," which are new cars plus used car sales weighted by a depreciation factor. He obtained an income elasticity of 1.8 to 2.0, and a price elasticity of −0.74 to −0.95. All of these can be thought of as long-run elasticities; on the basis of these figures, raising the price of automobiles would raise the total

2. Daniel B. Suits, "The Demand for New Automobiles in the United States 1929–1956," *Review of Economics and Statistics*, 40, no. 3 (August 1958): 273–280.
3. Gregory C. Chow, *Demand for Automobiles in the United States* (Amsterdam: North Holland Publishing Company, 1957).

amount spent on them. If they are truly long-run figures, they should be appropriate for estimating the equilibrium effect of a permanent price rise caused by pollution control costs.[4]

Dyckman[5] studied new car sales from 1929 to 1962, excluding 1942–1948 because of wartime restraints and associated dislocations. The dependent variable was the number of units sold; independent variables included discretionary income (real disposable income minus a subsistence amount) and new car prices adjusted by a special consumption deflator. The model was formulated in terms of first differences to avoid problems of changes in taste, but therefore measured only short-run responses to the variables. Credit and asset terms also were included. This model produced a price elasticity of demand for cars of −0.7, and an income elasticity of 1.7, using discretionary income. When real disposable income was used instead of discretionary income, the price and income elasticities became −0.8 and 4.0, respectively.

Other studies of demand for automobiles have been made by Houthakker and Taylor[6] and others[7] that suggest price elasticities in the order of −0.5 to −1.5 for various measures of purchase activity.

Long-run price elasticities of new car demand may differ substantially from short-run elasticities. The annual data used in most automobile demand studies shed far more light on short-run than on long-run demand. The Chow study, however, does not show the expected disparity, with its long-run elasticity of −0.74 to −0.95. In any event, existing evidence suggests a price elasticity in the vicinity of −0.8. Thus, a permanent price rise for automobiles would cause a somewhat less than proportional decrease in sales and a small rise in total expenditures for purchases.

The effect of such a price rise on total motoring is not easily identified. For many people, owning an automobile is a necessity, but purchasing a

4. The price data used by Chow are a simple average of all prices advertised in each of three newspapers on the last Sunday of the year. Since such data may not be properly representative of average annual prices, Chow's results are questionable.
5. Thomas R. Dyckman, "An Aggregate-Demand Model for Automobiles," *Journal of Business,* 38, no. 3 (July 1965):252–256.
6. H. S. Houthakker and Lester D. Taylor, *Consumer Demand in the United States, 1929–1970* (Cambridge, Mass.: Harvard University Press, 1966).
7. A summary of such studies is presented in Lawrence J. White, *The Automobile Industry Since 1945* (Cambridge, Mass.: Harvard University Press, 1971), p. 94.

new car is a luxury.[8] Thus, if new vehicles become relatively more expensive, there will be a substitution of maintenance for replacement, and the average lifetime mileage of the vehicle should increase. The size of the vehicle fleet and total vehicle miles, even in long-run equilibrium, will decline less than new car sales. This suggests that the elasticity of total motoring with respect to new car prices should be substantially less than one.

Suppose that society's objective is to reduce total motoring by 10 percent through an increase in the price of new automobiles. Assuming that motoring varies directly with fleet size, a drop in sales greater than 10 percent will have to be achieved in the long run. To reach the objective in the short run, there must be a still greater reduction in sales in the first years. With a price elasticity of demand equal to −0.8, a 10 percent sales reduction will require a price increase of 12.5 percent. A new car tax of 12.5 percent of retail value on annual sales of 10 million cars worth $30 billion will generate $3.75 billion of tax revenue. Lost consumer surplus will be

$$\frac{1}{2} \times dP \times dQ$$

or

$$\frac{1}{2} \times \$375 \times 1 \text{ million} = \$187.5 \text{ million.}$$

If all new vehicles were to have emission characteristics of 1971 General Motors cars, a reduction of 1 million new cars per year would reduce pollution by 0.45 million Pollution Units per year[9] at a cost of $950 of tax revenues per Unit and $47 per Unit of lost consumer surplus per year. Because motoring may fall less than sales, this represents a minimum cost.

Finally, increasing the price of new cars will be a slow means of reducing motoring. A price increase sufficient to reduce the rate of sales by 10 percent will reduce total motoring in the long run by 10 percent or less. In the first year, it will reduce total motoring by 1 percent or less, since cars one year old or newer comprise about 10 percent of the vehicle fleet.

8. White found that over 90 percent of new car buyers owned a previous automobile. Presumably most of these individuals could have retained their old vehicle, perhaps at an increased maintenance cost, instead of buying a new one. Ibid.
9. A Pollution Unit is the amount of pollution emitted by an average uncontrolled car in a year (10,000 miles).

While income is not a policy variable, income elasticities affect future vehicle sales. The income elasticities of demand for purchases have been found to be consistently high, in the 2 to 4 range. While these are usually short-run figures, household surveys generally show a positive income elasticity of ownership within the middle-income range.[10] If incomes continue to rise in the future, per-capita motoring also can be expected to increase if no cost changes occur. There may be a saturation point for new car purchases, as more and more families acquire two and three vehicles, but there is no evidence that this point is imminent.

Data on sales, fleet size, and vehicle mileage from 1945 to 1969 are presented in Table 6.1. It is notable that the ratio of annual miles driven to the size of the vehicle fleet has been relatively constant since 1945; automobiles were driven an average of almost 10,000 miles per year throughout the entire period. New car sales, however, varied widely; the number of automobiles per capita doubled; and real per capita income rose by 50 percent. If this constant miles per vehicle represents some underlying behavioral relationship rather than a fortuitous coincidence of diverse forces, then the effect of changing new car sales will be a proportional reduction in vehicle miles traveled, except as vehicle life increases. The data, however, are so collinear as to preclude a statistical test of the relationship. Later in this chapter, we will present tables showing that the various costs of motoring (such as gasoline, maintenance, and parking) appear to have risen more or less in proportion to one another—and this would tend not to alter vehicle life and utilization. Whether utilization could be changed by reasonable changes in the relative magnitude of purchase and operating costs is a question that must remain unanswered for the present.

An increase in the price of new cars will tend to reduce the percentage of households owning automobiles. But it may have the effect of increasing average mileage per vehicle, and it will also have adverse income distribution effects. Those most likely to give up ownership of an automobile are low-income households which can now barely afford it. Yet these are the people to whom automobiles are essential for transportation from home to work. Public transportation in most cities is oriented toward suburb-to-center-city work trips; but lower-income groups tend to live near the center

10. Automobile Manufacturers Association, Inc., *1970 Automobile Facts and Figures* (Detroit, Michigan, 1970), pp. 46, 47.

Table 6.1. Automobile Sales and Fleet Size, 1945–1969

Year	New Car Sales (millions of vehicles)		Auto Purchase Expenditures (billions of current dollars)	Registered Autos (millions of vehicles)	Annual Miles Driven (billions of miles)
	Domestic	Imports			
1945	0.07		0.36	25.8	200
1946	2.15		2.56	28.2	281
1947	3.56		4.84	30.8	300
1948	3.91		6.14	33.4	320
1949	5.12		8.64	36.5	342
1950	6.67	0.02	11.5	40.3	364
1951	5.34		10.1	42.7	392
1952	4.32		9.5	43.8	410
1953	6.12		12.8	46.4	435
1954	5.56		12.3	48.5	453
1955	7.92	0.06	16.8	52.1	493
1956	5.82	0.11	14.7	54.2	515
1957	6.11	0.26	16.3	55.9	529
1958	4.26	0.43	13.3	56.9	545
1959	5.59	0.67	17.2	59.5	573
1960	6.67	0.44	17.7	61.7	588
1961	5.54	0.28	16.0	63.4	605
1962	6.93	0.38	19.5	66.1	629
1963	7.64	0.41	21.5	69.1	645
1964	7.75	0.54	22.8	72.0	678
1965	9.31	0.56		75.3	709
1966	8.60	0.91		78.1	745
1967	7.44	1.02		80.4	766
1968	8.82	1.62		83.7	806
1969	8.22	1.85			

Sources: Automobile Manufacturers Association, Inc., *1967 Automobile Facts and Figures*, p. 5, and *1970 Automobile Facts and Figures* (Detroit, Michigan, 1970), pp. 3, 5; U.S. Department of Commerce, "Personal Consumption Expenditures by Type of Product," *Survey of Current Business*, 45, no. 11 (November 1965):22–23; and U.S. Department of Transportation, Bureau of Public Roads, *Highway Statistics Summary to 1965* (Washington, D.C.: U.S. Government Printing Office, 1968), Tables MV-200, VM-201, and *Highway Statistics 1968* (Washington, D.C.: U.S. Government Printing Office, 1968), Tables MV-1, VM-1.

city and work at distant points throughout the metropolitan areas—points that are often poorly served by the transit system.[11] Accordingly, it would seem more just to seek increases in marginal rather than average costs, thus permitting widespread automobile ownership but discouraging nonessential use of the car. A tax on operation rather than purchase would probably be the less regressive of the two, if the price elasticity of use were greater than the price elasticity of purchase for low-income groups, and if this were not true for higher-income groups.

Intercity Travel Demand
The demand for intercity motoring depends on the availability of substitute modes of transportation and the elasticity of demand for all intercity travel.

In considering the degree of substitution that can exist between the automobile and other intercity modes of transportation, we must first recognize that for the past two decades the automobile has dominated this travel. Since 1950, total expenditures on motoring, half of which may reasonably be attributed to intercity travel,[12] have been more than 20 times total expenditures on intercity railroad, bus, and airplane (see Table 6.2). Currently, 79 percent of all intercity trips are made by automobile, accounting for 86 percent of all intercity passenger miles. In nonautomobile travel, the rail share has been declining steadily, while the airline share has increased rapidly.

Since over 85 percent of intercity travel is by automobile, shifting a substantial number of motorists to other modes of transportation would cause a very large percentage change in the traffic volume of the beneficiary modes. Doubling the public transport share would cut vehicle miles by less than 20 percent, yet such a change would be a radical one indeed for the mass transportation modes. This change in utilization of public transport would

11. See Sumner Myers, "Personal Transportation for the Poor," Transportation and Poverty Conference, American Association for the Advancement of Science, Brookline, Mass., June 7–8, 1968.
12. The Department of Transportation shows automobile vehicle miles driven on urban streets as about 55 percent of urban plus rural automobile miles driven. The definition of urban streets and rural highways is not entirely satisfactory for the urban-intercity split, but should be a close proxy. U.S. Department of Transportation, Bureau of Public Roads, *Highway Statistics Summary to 1965* (Washington, D.C.: U.S. Government Printing Office, 1968).

Table 6.2. Intercity Travel by Mode, 1945–1969

Year	Purchased Intercity Transportation (millions of dollars)				Total Consumer Automobile Expenditures (millions of dollars) (Urban and Intercity)	Intercity Passenger Miles (millions of dollars)			
	Rail	Bus	Air	Total[a]		Rail	Bus	Air	Auto
1945	676	339	55	1107	3,990				
1946	567	338	82	1021	9,020				
1947	534	315	91	975	12,270				
1948	548	317	101	1000	14,670				
1949	466	319	116	932	17,910				
1950	394	309	141	872	21,870	32.5	26.4	10.1	438
1951	447	332	188	995	21,550				
1952	465	351	226	1070	22,040				
1953	442	346	266	1084	26,650				
1954	394	305	293	1026	26,770				
1955	378	295	349	1052	32,590	28.7	25.5	22.7	637
1956	385	304	394	1115	31,730				
1957	373	317	445	1167	34,760				
1958	338	296	479	1145	32,570				
1959	323	299	579	1233	37,980				
1960	319	313	646	1308	39,830	21.6	19.9	34.0	706
1961	309	321	707	1367	38,140				
1962	303	336	809	1478	42,520				
1963	268	316	857	1472	45,680				
1964	261	324	987	1603	47,950				
1965						17.6	23.8	58.1	818
1969						12.0	26.0	111.0	977

Sources: Automobile Manufacturers Association, Inc., *1970 Automobile Facts and Figures* (Detroit, Michigan, 1970), p. 54; and U.S. Department of Commerce, "Personal Consumption Expenditure by Type of Product," *Survey of Current Business,* 45, no. 11 (November 1965):22–23.

a. The total includes other miscellaneous modes which amount to 2–3 percent of the total.

require a large incentive, such as substantial price reductions and/or service improvements; either alternative would be expensive. It would also require a great capital investment for new vehicles and associated facilities,[13] and therefore could not be accomplished in a short period of time.

In the absence of any trend toward the public modes of transportation, some positive action will be required to increase their share. This might take the form of a permanent subsidy to change relative prices, a research and development program to induce technical progress that improves cost/performance, or restrictive taxes on automobile use.

The airline share of intercity traffic has grown at the expense of the railroads and more rapidly, in percentage terms, than automobile traffic; the cost savings of the jumbo jets make further growth likely. Even this very rapid expansion, however, has not reduced automobile passenger-miles, although it may have slowed the rate of growth somewhat. Moreover, if airlines could significantly reduce the amount of intercity motoring, it is still not clear that airplanes produce less pollution per passenger where they take off and land than automobiles in the urban areas.[14]

It is sometimes said that better railroad service would attract passengers from both automobiles and airlines, reducing the external effects of these two modes. This suggestion ignores the fact that, for the nation as a whole, rail passenger-miles comprise less than 2 percent of automobile passenger-miles and 10 percent of airline passenger-miles. Doubling or tripling rail travel would have little impact on the other modes in the aggregate. In specific corridors where rail travel is already large, such as the Boston-Washington corridor, an increase in the rail share may be significant; but there are few other instances where this is possible. Furthermore, there is no evidence that improved rail service could be self-supporting, except in a few heavily populated corridors, so that the implicit cost of any solution that promotes rail travel will probably be quite high. Thus improved rail

13. Recent experience with rail passenger service, generally, and high-speed rail service, in particular, is evidence of this.
14. For a discussion of aircraft emissions, see "Air Pollution Control Research into Fuels and Motor Vehicles," Hearings before the Subcommittee on Public Health and Welfare of the Committee on Interstate and Foreign Commerce, House of Representatives, 91st Congress, June 19, 1969, Appendix A. In flight, airplanes emit about one-tenth as much pollution per pound of fuel as an uncontrolled automobile engine, but they emit substantially more at low levels during ground operations, takeoff, and landing.

service (if it is ever feasible) can aid only in local, not in national, pollution reduction.

One study of intercity travel in the Northeast Corridor has investigated elasticity of demand for travel by automobile and other modes in the Washington-Boston corridor.[15] It was found that the cost elasticity of demand for automobile travel for personal trips was −0.93, while for business trips it was −0.36. Automobile travel was more sensitive to travel time, with a time elasticity of −1.36 for personal trips and −3.4 for business trips. The low cost-elasticities of demand in this corridor—where rail, bus, and air service is probably better than anywhere else in the country—suggest that heavy taxes would have to be placed on motoring to effect a substantial reduction in automobile travel. If −0.6 is taken as an average price elasticity of demand for intercity motoring, and $0.05 per mile is used as the marginal cost of operating an auto, a 10 percent reduction in intercity motoring would require a marginal cost increase of almost 17 percent, or $0.0085 per mile. For the country as a whole, a tax in this amount on intercity travel would generate $3.4 billion of revenue per year, for a reduction in pollution of 3 million Pollution Units. The lost consumer surplus would be

$$\tfrac{1}{2}\, dP \times dQ = \tfrac{1}{2} \times \$0.0085 \times 40 \times 10^{9}$$
$$= \$170 \text{ million per year.}$$

This is a tax revenue of $1120 per Pollution Unit abated, and a loss in consumer surplus of $56.60 per Pollution Unit abated in intercity travel. Slowing the average speed of intercity motoring would have a stronger effect, but this is hard to do in practice. Reducing highway capacity to increase congestion would hit peak period travelers more than off-peak, would affect buses and trucks as well as automobiles, and might increase pollution per vehicle mile. Reducing speed limits well below the design speed of the highway would probably cause very large enforcement costs and again would affect buses and trucks as well. In short, automobile travel time is not a good short-run policy variable, especially if a reduction is desired.

In the long run, intercity motoring may be reduced by the diversion of investment funds from intercity highway construction to facilities for other

15. Systems Analysis and Research Corporation, "Demand for Intercity Passenger Travel in the Washington-Boston Corridor," prepared for the Office of the Under Secretary of Commerce for Transportation, U.S. Department of Commerce, 1965.

modes of transportation. In the short run, only large cost increases are likely to have much effect on aggregate motoring, and these will be regressive since the rich tend to spend a smaller percentage of their income on automobiles than do middle- and low-income households. Encouragement of substitute modes of travel, such as railroad and bus, could have an impact in areas where their share is already large; but for the country as a whole, this method can have little impact.

Urban Travel Demand

In terms of total contribution to air pollution, urban motoring is probably as much to blame as intercity motoring, since total vehicle mileage in each category is about equal. In terms of the impact of the pollution, however, urban motoring is the greater offender, because the number of persons exposed to a given unit of pollution is much higher and the average air quality is worse, adding to the marginal damage per unit (assuming that marginal harm increases with pollution density). The strongest proposals for reduction in motoring have referred to the urban context and the need to control urban air pollution.

Schemes to reduce urban motoring by more extensive use of mass transit are confronted by the same problem encountered in intercity travel: mass transit accounts for only a small portion of total urban passenger movement. Table 6.3 shows that expenditures on all forms of local transportation except private automobiles (but including taxicabs) are only 4 percent of the expenditures on automobiles. Moreover, if half the automobile expenditures are urban, automobiles still account for 90 percent of urban expenditures. Public transit is used more for work trips than for any other type of urban trip, yet only 14 percent of all work trips are made by public transit.[16] It seems likely that today less than 10 percent of all urban passenger-miles are traveled in public transit. Thus even a large growth in transit usage—say, a doubling—would have a small impact on urban motoring, reducing it by 11 percent. Moreover, the trend in transit usage has been steadily downward, while automobile use has increased rapidly.[17]

16. Automobile Manufacturers Association, Inc., *1970 Automobile Facts and Figures*, p. 53.
17. For an analysis of the reasons for a downward trend in the use of one mode of public transportation, see Donald N. Dewees, "The Decline of the American Street Railways," *Traffic Quarterly*, October 1970, pp. 563–581.

Table 6.3. Urban Travel by Mode, 1945–1967

Year	Purchased Local Transportation (millions of dollars)				Total Consumer Automobile Expenditures (millions of dollars) (Urban and Intercity)	Passenger Travel	
	Bus, Subway, Street-car	Taxi	Com-muter Rail	Total		Total Transit Trips (subway, bus, streetcar) (billions of passenger trips)	Urban Auto Vehicle Miles (billions)
1945	1316	372	58	1746	3,990	23.3	100
1946	1334	511	63	1908	9,020		
1947	1328	532	67	1927	12,270		
1948	1405	508	76	1989	14,670		
1949	1407	465	79	1951	17,910		
1950	1368	487	79	1934	21,870	17.2	182
1951	1371	511	83	1965	21,550		
1952	1382	519	89	1990	22,040		
1953	1385	530	93	2008	26,650		
1954	1333	509	96	1938	26,770		
1955	1292	540	101	1933	32,590	11.5	246
1956	1285	579	107	1971	31,730		
1957	1255	616	116	1987	34,760		
1958	1219	574	124	1917	32,570		
1959	1244	602	125	1971	37,980		
1960	1270	609	122	2001	39,830	9.4	294
1961	1256	570	127	1953	38,140		
1962	1266	588	127	1981	42,520		
1963	1251	595	130	1976	45,680		
1964	1271	593	134	1998	47,950		
1965						8.3	353
1966							
1967						8.2	383

Sources: American Transit Association, *1968 Transit Fact Book* (Washington, D.C., 1968), p. 57; and U.S. Department of Commerce, "Personal Consumption Expenditure by Type of Product," *Survey of Current Business,* 45, no. 11 (November 1965):22–23.

Moses and Williamson[18] have studied the price changes necessary to divert motorists from automobiles to mass transit in Chicago. They conclude that up to 77 percent of automobile commuters (work trip only) could be diverted to some other mode if the price of automobile commuting were raised by $1 per trip or $2 per day. For a ten-mile, one-way trip, this would require a doubling of average per-mile automobile cost or a large increase in parking fees. If one were to reduce transit prices instead of increasing automobile costs, a *negative* transit price would be necessary to attract 50 percent of motorists from their cars. If transit were made free, less than one-fifth of the motorists would be diverted. These figures, furthermore, are for commuters only; such policies would have a lesser impact on shoppers and other travelers. Given that the study applied only to the work trip, however, and only to those who had a transit alternative,[19] it is clear that the policies studied could reduce total urban motoring only by a small percentage.

These conclusions have been corroborated and extended in other studies. Domencich and Kraft[20] conclude that reducing the price of mass transit to zero will increase ridership only by one-sixth to one-third, having a negligible effect on motoring.[21] Moreover, Domencich, Kraft, and Valette[22] have suggested that reasonable improvements in transit travel times will not seriously reduce total motoring. This is not inconsistent with findings that improved service can bring substantial increases in transit ridership; since transit is only 10 percent of total urban travel, a 40 percent increase in transit passengers would reduce motoring by only 4 percentage points if *all*

18. Leon N. Moses and Harold F. Williamson, Jr., "Value of Time, Choice of Mode, and the Subsidy Issue in Urban Transportation," *Journal of Political Economy*, 71, no. 3 (June 1963):247–264.
19. That is, those who worked predominantly in the central business district. Most transit trips are CBD-oriented, while such automobile trips are only a small portion of total motoring in any urban area.
20. Thomas A. Domencich and Gerald Kraft, *Free Transit* (Lexington, Mass.: Lexington Books, 1970).
21. They estimate that free transit in Boston, which is one of the few U.S. cities with a subway system, will reduce automobile work trips by 14 percent and total automobile trips in the area by 4 percent. Ibid. p. 85.
22. Thomas A. Domencich, Gerald Kraft, and Jean-Paul Valette, "Estimation of Urban Passenger Travel Behavior: An Economic Demand Model," presented at the Annual Highway Research Board Meeting, January 1968 (Cambridge, Mass.: Charles River Associates, 1968).

of the added travelers were previously motorists, which some generally are not.[23]

If urban motoring is to be reduced, it will generally be necessary to restrict auto usage while increasing its price, and to improve mass transit facilities and service. Sweden has successfully reduced motoring, particularly in Stockholm, by closing portions of some streets entirely, limiting portions of other streets to buses and taxis, providing exclusive bus lanes downtown, limiting street parking, charging high off-street parking fees, and providing high-quality bus, subway, and commuter railroad service. This program provides some hope for U.S. cities prepared to take such comprehensive and rather revolutionary steps.

One way to evaluate reductions in motoring as a means of pollution control is to examine recent public transit expansions, allocating net project cost to the automobile trips diverted. For example, the very successful Lindenwold High Speed Line in Philadelphia and New Jersey meets its operating expenses from fares collected, but a $94 million investment, with debt service of $5.4 million per year, is not paid by passengers.[24] With 33,000 passengers per day, 40 percent of whom previously drove automobiles (assuming an average ten-mile trip), this is 133,000 vehicle-miles per day of motoring reduction, representing 13.3 Pollution Units from uncontrolled cars. The debt service is $18,000 per day (excluding Sundays); thus, if the debt service represents the net social cost of the project and can all be allocated to pollution control, it costs about $1,400 per Pollution Unit abated. This is over ten times the cost of recent and proposed new car programs.[25] Of course, if other objectives, such as reducing highway congestion, are

23. The experiments on which service change impact is generally based involve a single line or mode. Some new passengers come from automobiles, some from other transit, and some are making trips not previously made. The per-passenger cost of such changes is generally quite large. Wheeler has studied service and fare experiments on the Boston and Maine commuter service in Boston. He concludes that peak-period users are not responsive to service improvements and off-peak users are moderately responsive, but that no amount of changes in fares and schedules can eliminate the large operating deficit, much less cover capital costs. The *average* deficit for this commuter service is about $100 per passenger per year, or just under $0.04 per mile. This implies an abatement cost of $400 per Unit, well above the cost of most other means of pollution reduction. The marginal deficit for rush-hour service may be even larger. See Porter K. Wheeler, "Price and Service Sensitivity in Urban Passenger Transportation," Ph.D. Thesis, Department of Economics, Harvard University, 1968.
24. Charles Burch, "The Little Railroad That Could," *Fortune,* July 1971, p. 72.
25. See Chapter 7.

major justifications for the project, then only a portion of the deficit need be allocated to pollution control, and its cost effectiveness will be correspondingly better.

In contrast to the Lindenwold Line, the Skokie Swift Transit Line in Chicago during its first two years covered operating expenses and had reserve enough almost to carry the debt service on investment.[26] Thus, the cost of eliminating 0.75 Pollution Units per day there is zero. Even in cases such as the Skokie Swift, however, where mass transportation expansion is a highly cost-effective way to reduce pollution, the reduction will generally be such a small portion of total motoring in the metropolitan area that it cannot constitute a policy for solving the area's pollution problems.

The above data suggest that in the aggregate, whatever the price elasticity of demand for urban motoring, the price elasticity of demand for mass transit is low and irrelevant, since transit use is a small percentage of total urban movement. The service elasticities are larger, but the quality of service is a difficult policy variable to use and improvements may be prohibitively expensive. In a few cities, such as New York, where mass transportation carries a large portion of the total travelers, it may make sense to try to encourage transit and discourage motorists through price changes or constraints on a city-wide basis. Elsewhere, trips to a particular area may be affected—for example, to the central business district, where transit usage already is high—and pollution may thus be reduced *in that portion of the city*. It must be recognized, however, that only a comprehensive and city-wide program can have much impact on the total emissions and average concentration for the entire city. Promoting transit usage will have little effect on total emissions unless the automobile is simultaneously restricted.

Time, Cost Structure, and Price Elasticity of Demand for Motoring
We have talked rather glibly about the price elasticity of demand for motoring, as if each vehicle-mile were like any homogeneous commodity that can be purchased in any amount for a given price per unit. In fact, several characteristics of motoring distinguish it from other goods, which helps to explain some of the empirical results found thus far and suggest what parameters will be feasible policy instruments.

26. Chicago Transit Authority, "Skokie Swift The Commuter's Friend," Mass Transportation Demonstration Project Final Report, May 1968.

A major element in motoring decisions is the fact that the monetary cost of motoring consists of a fixed and a variable portion; in both cases, perceived costs and actual costs may differ, since neither is paid for on a per-mile basis. The average cost of motoring has frequently been estimated at about $0.10 per mile, while marginal cost of an incremental trip has been estimated at anything from that value down to a few cents.[27] A recent and reliable estimate of motoring costs has produced an estimated total cost of $0.1189 per mile, for a 1970 vehicle, at 1970 prices.[28] The breakdown of these costs is shown in Table 6.4. These costs are derived by multiplying

Table 6.4. Cost of Operating a Typical Automobile (1970 vehicle, 1970 prices)

	Average Cost (cents per mile)
Items Excluding Taxes	
Depreciation	3.22
Gasoline	1.73
Repairs and Maintenance	1.52
Replacement Tires	0.39
Oil	0.16
Insurance	1.72
Home Garage	1.20
Parking and Tolls	0.60
	10.54
Taxes and Fees	
Registration	0.20
Titling	0.13
Gasoline	0.80
Auto and Tires	0.22
	1.35
Subtotal	11.89
Opportunity Cost of Capital	0.44
Total	12.33

Source: E. M. Cope and C. L. Gauthier, "Cost of Operating an Automobile," U.S. Department of Transportation, Federal Highway Administration, Bureau of Public Roads, February 1970, p. 10.

27. Meyer, Kain, and Wohl base their computations for commuter costs on a special-purpose commuting car, perhaps like current subcompacts, and arrive at a marginal cost of $0.029 per mile, in 1964 prices, for gas, oil, tires, and maintenance. See John R. Meyer, John F. Kain, and Martin Wohl, *The Urban Transportation Problem* (Cambridge, Mass.: Harvard University Press, 1965), p. 218.
28. E. M. Cope and C. L. Gauthier, "Cost of Operating an Automobile," U.S. Department of Transportation, Federal Highway Administration, Bureau of Public Roads, February 1970, p. 10.

the actual typical quantities purchased in each category by the average price of the items. The vehicle is assumed to last ten years and to be driven 10,000 miles per year.[29] The only item omitted from the original study is the cost of capital value that is tied up in the vehicle. This is estimated by multiplying the market value of the vehicle at the end of each year by 0.06, representing a 6 percent opportunity cost of capital, and dividing by the number of miles driven. This produces an average cost of $44 per year, or $0.0044 per mile.

The costs in Table 6.4 are built up from assumptions about a particular "typical" vehicle. To check these costs, the aggregate annual expenditures made by U.S. motorists are estimated and then converted into per-mile figures. The results of these calculations for 1968 are shown in Table 6.5. The costs of depreciation, gasoline, maintenance, tires, and oil are quite similar to those of Table 6.4; the remaining items and the total are signifi-

Table 6.5. Total Cost of Motoring, 1968

	Billions	Cents per Mile
Depreciation	$25.4	3.15
Gasoline	20.0	2.48
Maintenance	10.9	1.35
Tires	4.3	0.53
Oil	1.2	0.15
Insurance	7.2	0.89
Tolls and Parking	1.3	0.16
Total	$70.3	8.71

29. Recent research has shown the average age of automobiles at scrapping in the United States to be just under ten years. (See J. A. Fay and M. Scott Mingledorff, "Dynamics of Automobile Population and Usage," Working Paper no. 2, prepared for the Legislative Research Council of Columbia University Conference on the Auto and the Environment, April 18 and 19, 1972, Columbia University, New York.) This figure is reinforced by the Automobile Manufacturers Association finding that the average age of the fleet is just over five years and has not changed significantly since the early 1950s. This clearly refutes the conventional wisdom that modern cars fall apart in a few years, and that they are less durable than their predecessors. Combined with other data showing annual average mileage of almost 10,000 miles, it implies an average lifetime of about 100,000 miles. This mileage is accumulated fairly rapidly in early years then more slowly. In the first year, about 13,500 miles may be accumulated; in the fifth year, there will be about 10,000, and in the tenth year, less than 7,000. (See Thurley A. Bostick and Helen J. Greenhalgh, "Relationship of Passenger-Car Age and Other Factors to Miles Driven," Highway Research Record no. 197, Highway Research Board, Washington, D.C., 1967, p. 33.)

cantly lower. The aggregate statistics do not impute a value to home parking space, or include home garaging fees where these are paid separately
from rent, and thus are underestimated. Taxes, aside from the gasoline tax,
are not included in the aggregate study. The insurance costs are also inex-
plicably lower. It appears that most of the typical costs are confirmed by the
aggregate statistics. Where there are differences, the former are more accurate, because of reporting omissions in the aggregate data. Table 6.4 will
therefore be used in the following analysis.

The true marginal cost of an incremental mile in an existing vehicle
should be the sum of those costs which are a function of mileage rather
than time. Within a broad range of mileage accumulation rates, depreciation is not affected by actual miles driven; thus, one-quarter of depreciation
will be deemed variable. Although oil change intervals are specified in
terms of mileage or time, whichever comes first, the oil cost is considered
variable with mileage; few motorists actually follow the recommended
schedule. Tires, maintenance, tolls, and nonhome garage parking expenses
are all directly related to mileage driven. Insurance claims for all but theft
and acts of God must be more or less proportional to exposure to accidents
—that is, mileage driven. Since the true marginal cost of motoring should
include a large portion of accident costs, we will consider three-quarters of
insurance costs to be variable with mileage. The true marginal cost of motoring totals $0.0729 per mile, as shown in Table 6.6.

Table 6.6. Private Marginal Cost of Motoring
(1970 vehicle, 1970 prices)

	Marginal cost (cents per mile)
1/4 Depreciation	0.80
Gasoline	1.73
Gasoline Tax	0.80
Repairs and Maintenance	1.52
Tires	0.39
Oil	0.16
3/4 Insurance	1.29
Parking and Tolls	0.60
Total	7.29

Source: Table 6.4.

It seems quite clear, however, that this is not the cost to which most persons considering a trip in an available vehicle respond. The majority of motorists do not know their maintenance and tire costs, and do not see insurance as varying with mileage (for the individual, it does not). When people speak of the cost of a trip, they usually mention gasoline, tolls, and parking—and if it is a particularly old car, oil as well. As Table 6.7 shows, these total $0.0329 per vehicle-mile. These elements appear to be the maximum perceived dollar marginal cost; they may even be overestimated, because they need only be paid at the time of the trip if a long trip is taken.

For many automobile trips, however, none of the above figures may represent adequately the marginal cost of the trip. Since the demand for transportation is usually a derived demand arising from the desire to be at a certain place at a certain time, the factors a person considers when weighing trip decisions include not only monetary costs but other travel parameters, such as waiting and travel time, variance of time, and comfort. To the extent that the nonmonetary costs of a trip can be quantified, they should be included in any analysis of transportation decision making.

Some research has been done on the value of time to motorists.[30] The best of the studies have produced estimates of $0.75 and $2.74 per passenger-hour, and $1.02, $1.42, and $2.94 per vehicle-hour. These suggest a value somewhere in the $1 to $3 per vehicle-hour range, or $0.65 to $2 per

Table 6.7. Out-of-Pocket Motoring Costs
(1970 vehicle, 1970 prices)

	Cost (cents per mile)
Gasoline	1.73
Gasoline Tax	0.80
Oil	0.16
Parking and Tolls	0.60
Total	3.29

Source: Table 6.4.

30. See Meyer, Kain, and Wohl, *Urban Transportation Problem*, pp. 102–104; Moses and Williamson, "Value of Time"; and James R. Nelson, "The Value of Travel Time" in Samuel B. Chase, Jr., ed., *Problems in Public Expenditure Analysis* (Washington, D.C.: The Brookings Institution, 1966).

passenger-hour. Allowing for inflation of about 25 percent since 1960, here we will use a value of $2 per passenger-hour.

The value of time to travelers and shippers can be explicitly incorporated into a formal decision model. A model has been developed in which the total cost of a trip is the sum of monetary and nonmonetary costs, where the latter include the value of both waiting and travel time.[31] This total cost is referred to as the R-factor cost, where R stands for resistance to movement. It is hypothesized that it is this R-factor cost to which travelers respond in making modal choice and whether-to-travel-or-not decisions. If we consider just monetary cost and time, then the R-factor of a trip is

$$R = C + T \times V, \tag{6.1}$$

where R = total cost, C = monetary cost, T = time needed, and V = value of time in dollars per hour.

Let us examine urban travel decisions using the R-factor instead of simple monetary costs. In urban areas, the average vehicle is assumed to carry 1.7 passengers and to move at 22 miles per hour.[32] Under these conditions, the average value of time is $0.15 per mile. The perceived R-factor cost is the perceived monetary cost plus the time cost, or $0.18 per mile. For intercity trips, speeds may average 50 miles per hour, and there are often 2.5 persons per vehicle; so that the value of time would be $0.10 per mile, or less if intercity travelers value their time at a lower rate.

It is now apparent why studies of the demand for motoring show a low price elasticity: the monetary cost is only a small portion of the total trip cost. To raise the marginal R-factor cost of a trip by 10 percent will require a cost increase of $8 to $15 *billion* per year. If this cost increase were to

31. Paul O. Roberts, David T. Kresge, and John R. Meyer, "An Analysis of Investment Alternatives in the Colombian Transport System," final report, Transport Research Program, Harvard University, 1968, Chap. 4. For other models which formally incorporate time as a decision parameter, see Northeast Corridor Transportation Project, "Approaches to the Modal Split," Office of the Under Secretary of Commerce for Transportation, U.S. Department of Commerce, Technical Paper no. 7, February 1967; and Reuben Gronau, "The Effect of Traveling Time on the Demand for Passenger Transportation," *Journal of Political Economy*, 28, no. 2 (March–April 1970): 377. See also Donald N. Dewees, "The Impact of Urban Transportation Investment on Land Value," University of Toronto–York University Joint Program in Transportation, Research Report no. 11, April 1973.
32. Meyer, Kain, and Wohl, *Urban Transportation Problem*, pp. 103–234.

come from an increase in the gasoline tax (of $0.15 to $0.25 per gallon), its primary impact would be on choice of vehicle, not on miles traveled, shifting buyers to more economical models. It is not clear that this would reduce emissions at all, since cars of all sizes must now meet the same emission standards in grams per vehicle-mile.[33]

Instead of raising the gasoline tax, one may raise or levy tolls on automobile use to reduce total motoring. This can be done easily on existing toll highways and at moderate expense on currently free limited-access superhighways. The technology does not exist at present to make it possible to collect tolls for the use of local streets and roads at a reasonable cost. Ground-based devices are out of the question, because the hundreds of thousands of miles of local streets and roads would require a great many installations. Vehicle-mounted devices, such as a sealed odometer that would be read once a year, might be relatively inexpensive, but also could be subject to tampering by the motorist. Only after extensive field experience can it be decided whether a meter could be built that is tamperproof. A superhighway toll will reduce travel in specific corridors but will not have much impact on total travel in a metropolitan area, since most travel is on local streets, and some motorists may avoid tolls by using parallel highways; a meter toll will tend to reduce all motoring without discrimination between heavily polluted and relatively clean areas. The choice between the two thus depends upon whether one is concerned with local pollution peaks or total emissions in a city or state.

As with the gasoline tax, the toll rate would have to be high to effect even a 10 percent reduction in motoring.

As a low estimate, suppose that the elasticity of demand for all motoring is -1 with respect to total R-factor marginal costs, $0.18 per mile. Then a 10 percent reduction in motoring will require an increase in marginal cost of $0.018 per mile. This will generate revenues of $14.4 billion per year and cause a loss of consumer surplus of

$$\tfrac{1}{2} \times 0.018 \times 80 \text{ billion} = \$720 \text{ million per year.}$$

33. One study of automobiles with exhaust emission controls in California shows no significant correlation between engine size and pollution emission in grams per mile. See R. d'Arge, T. Clark, and O. Bubik, "Automobile Exhaust Emissions Taxes: Methodology and Some Preliminary Tests," Project Clean Air, Research Project S-12, University of California at Riverside, September 1, 1970.

With pollution reduced 6 million Pollution Units, this is $2,400 of revenue per Pollution Unit and $120 of lost consumer surplus. As a high estimate, suppose that the elasticity of demand for all motoring is -1 with respect to perceived operating cost, $0.033 per mile. Then a 10 percent reduction in motoring will require an increase in marginal cost of $0.0033 per mile. This will raise $2.4 billion per year, with a loss of consumer surplus of

$$\tfrac{1}{2} \times 0.0033 \times 80 \text{ billion} = \$132 \text{ million.}$$

This is $440 of revenue per Pollution Unit, and $26 of lost consumer surplus. It is likely that the time cost of a price strategy to reduce motoring would be between these extremes. If motoring is currently excessive for other reasons, such as causing congestion, then these losses in consumer surplus may be more than offset by social gains from reducing congestion.

Parking fees may be increased in an effort to discourage motoring to a particular area. This is feasible only where most parking is in lots or garages, so that fees can be collected efficiently. In areas where much of the parking is on the street, enforcement of meter violations may become expensive if high rates cause widespread violations. Even in garage-dominated areas, high rates may result in general disregard for street parking limitations or prohibitions, and attendant high enforcement costs. If high parking charges can be collected, they are a powerful tool for discouraging driving to particular destinations. They may therefore reduce pollution in specific congested areas. Unless they can be imposed upon an entire metropolitan area, however, they cannot hope to achieve a significant reduction in total motoring. The clear inferiority of most automobile substitutes means that restricted motoring to a particular area will probably reduce total travel to that area, unless the aesthetic gains from reduced traffic are substantial. Thus few areas might consent to increased parking fees, unless an adequate alternative transit service were supplied. Parking fees are consequently a selective and perhaps discriminatory tool for reducing total motoring.

Finally, motoring may be reduced if costs that are now fixed (varying with time) are increased on a mileage basis. One candidate for such treatment is insurance, part of which is actually related to mileage driven. If a tamperproof odometer can be built, then part of the insurance premium (say, collision and liability) can be levied as so much per mile rather than so much per year. A $100 per year premium will become $0.01 per mile for the average 10,000 miles per year driven. This device would achieve, perhaps, a 6 to 10 percent increase in R-factor cost, a 30 percent increase in perceived monetary operating cost, and thus probably a 10 percent decrease in annual motoring with negligible income transfer consequences (from high-mileage drivers to low-mileage drivers). Those who use their car sparingly will be better off than before. If the technology of a tamperproof odometer can be perfected, this would seem to be a very attractive strategy to rationalize motoring decisions and reduce total motoring.

In computing the cost of motoring reduction, we have implicitly assumed that the amount of motoring is optimal except for pollution. In urban areas, however, automobiles cause highway congestion, slowing traffic flow. Congestion is an externality like pollution in that the individual motorist does not perceive the full social cost of his driving, including the extent to which his car slows others. Thus this congestion alone would justify some additional cost per mile, or congestion toll, which some authors have estimated as high as $0.10 per mile. This would significantly reduce motoring. In the absence of such a toll, the amount of motoring is clearly much more than optimal. Thus, where motoring is seriously underpriced because of congestion, imposition of motoring reduction programs would have benefits greater than calculated here; in some cases many times greater. Despite the difficulty of achieving motoring reductions, they may be highly desirable in many urban areas.

Conclusions
The preceding review and analysis has led to several important conclusions regarding reductions in motoring as a means of reducing automobile emissions. The long-run price elasticity of demand for new car sales seems to be less than one. Thus, in the long run, raising the average price of new cars will cause a less than proportional decrease in sales; and unless maintenance and operating cost are raised simultaneously, the elasticity of total motoring with respect to new car prices may be small indeed. Raising the price of new cars will have little short-run impact, since it will be years before a reduced rate of new sales has its effect on the total size of the vehicle fleet. Furthermore, a tax-induced price change will be regressive, taking a greater

percentage of income from lower-income than higher-income groups.[34] These disadvantages seem high for a relatively impotent policy.

Increasing marginal costs of all motoring is a difficult means of reducing automobile pollution, requiring very large price increases. Marginal monetary operating costs are a small part of the total cost of most trips and are poorly perceived, so that very large taxes will be needed to achieve a significant effect on total motoring. If a tax is imposed on vehicle operation rather than on sales, it will cost $400–$2,400 of tax revenue per Pollution Unit reduction and $24–$120 per Unit of lost consumer surplus (not counting congestion benefits), which brackets the performance of increasing new car prices. Charging tolls for road use also cannot have much of an effect on total emissions. Higher tolls on existing toll roads will discourage only a small percentage of motorists (given observed price elasticities of demand for motoring), and some of them will simply use slower, parallel toll-free roads. Increased parking fees may discourage driving to particular areas where most parking is paid for, reducing local emissions significantly if high-quality public transportation service to the area is available.

One case where raising marginal cost is particularly attractive is in any shift of annual or fixed costs to a mileage rate. This will not reduce the availability of automobiles to those who need them, but will reduce total mileage driven. Such a change is virtually costless and can have a significant effect if major costs, such as insurance, are shifted to the new rating scale.

The use of public versus private transportation depends on land-use patterns, the spatial arrangement of the city, relative costs, quality of service, and consumer preferences. Public transit carries only a small portion (10 to 15 percent) of total local and intercity travel. Even if large amounts were spent on public transportation, it would probably take years to double its share of the total transportation market. The costs, both direct and indirect, would be enormous, and the percentage reduction in motoring small unless auto use is simultaneously restricted. Substitution of transit for automobile travel seems to be a feasible means of pollution reduction primarily in cities where there is still a heavy reliance on public transport; even there, its main impact will be confined to limited areas, such as the central business district. In the long run, changes in the form of cities may reduce the

34. The absolute impact will be greater on the rich, but a neutral tax here is deemed one that takes a constant percentage of all incomes, not a constant absolute amount.

amount of automobile travel demanded, but this is a hope for new towns, or for old towns decades from now—it has little promise in the short run. In addition, such changes are counter to all recent trends in urban land use and transportation. The significant gains in automobile pollution control for all but small areas of large cities must be made in reducing pollution per vehicle-mile, not in reducing vehicle-miles themselves, unless we are prepared to accept serious restrictions on driving or large price increases for motoring.

Cost and Effectiveness of Technical Alternatives

7

Choosing among pollution abatement strategies requires some understanding of the relative performance of the technical means of abatement available. If perfect information were available, it would be possible to perform a cost-benefit analysis of alternative strategies and devices to determine the optimum combination. The problems of estimating benefits (developed in Chapter 3), however, preclude such analysis, leaving cost effectiveness as the next-best alternative. This chapter assesses the cost and effectiveness of a variety of technological alternatives that change the joint production function for automobile transportation and air pollution by modifying the engine or the fuel or by treating the exhaust gas. All costs and effects are computed on a per-vehicle basis; changes in total motoring, which were considered in Chapter 6, are not included.

The purpose of these calculations is to illustrate which avenues of control are most fruitful and what degree of abatement can be achieved at what cost. This permits construction of a marginal cost-of-abatement curve. The results cannot be interpreted as the basis for deciding precisely what devices to choose; because the data are imprecise and the technology is advancing rapidly, our results can rapidly become obsolete. Thus they can be taken to recommend or condemn specific devices or degrees of abatement only when they are shown to represent accurately the cost and performance attainable at the time of the decision.[1]

The systems analyzed here all represent technology now in use or in the testing and prototype stage of development. Almost half of the devices have already been included in domestic automobiles. The analysis thus does not look into the future much beyond the mid-1970s. As other devices are developed and cost and performance data for them become available, evaluations similar to those here can be made.

The details of cost and effectiveness calculations are presented in Appendix C. Clearly dominated systems are not included, since only the frontier of the production function is of interest. Most systems are considered for new automobiles only, on the grounds that the abatement per dollar that can be achieved on these vehicles is generally well above that for used cars. For comparison purposes, three used-car systems are included.

---

1. For an extensive review of automobile pollution controls, see Joe S. Bain, "The Technology, Economics, and Industrial Strategy of Automotive Air Pollution Control," *Western Economic Journal*, 8, no. 4 (December 1970):329–356.

All of the costs used in this study, and particularly those for prospective control systems, are subject to some uncertainty and possible error. In order to provide a simple means for evaluating the importance of possible error and for making quick estimates of the cost of devices not included here, we have performed a sensitivity analysis. In this analysis a base situation—the uncontrolled vehicle—is subjected to small variations in factor costs and proportions. The impact of these changes on total cost is noted in terms of capital cost, operating cost, and total cost. More detailed analysis can be made of individual control systems to determine how important each input is to their final cost. The results of the analysis are presented in Table 7.1.

Automobiles are assumed to be homogeneous, with specifications about the average of the 1970 U.S. fleet: curb weight is 3,600 pounds, and the average power is 230 HP.[2] In fact, the impact on individual vehicles will

Table 7.1. Sensitivity of Total Cost to Parameter and Factor Price Changes

Factor or Parameter	Base Value	Change			Resulting Cost Change		
		Absolute Value	As a Percentage of Base (%)	Capital ($)	Variable per Mile ($/mile)	Total per Mile ($/mile)
Fuel Economy	15 MPG	− 1.5	−10	—	+0.0026 (+0.0017)	+0.0026 (+0.0017)
Fuel Price	$0.35/gal ($0.24/gal)	+$0.035 (+$0.024)	+10 (+10)	—	+0.0023 (+0.0016)	+0.0023 (+0.0016)
Power Price	$1.69/HP	$0.17/HP	+10	+ 39.00	—	+0.0004
Capital Cost	$3,000.00	$300.00	+10	+300.00	—	0.0041
Annual Maintenance	—	$10.00	—	—	+0.001	0.001
Vehicle Life	10 years	−1 year	−10	10.00	—	0.00011 per $100.00

Note: Numbers in parentheses are resource cost figures, from which the fuel tax has been omitted.

2. Average power of domestic new cars was about 245 horsepower in 1968 (see Appendixes A and B). Foreign cars represented about 10 percent of the U.S. market and probably averaged about 100 horsepower. This would give an average fleet horsepower of about 230.

deviate from the average because of differences in power, weight, and other parameters.

Abatement Measurement

The comparison of one abatement device with another is made difficult by the fact that there is not a single pollutant but four, and their effects and relationships are not well understood. A single index is needed to measure the change in emissions caused by an abatement device and to allow comparisons of effectiveness. Here we will use a simple statement of the emission rate for each pollutant in grams per mile and the five indices that were developed in Chapter 3. Most comparisons will use Net Percent and the Pollution Unit, which are equivalent except for a scale factor of 100. It may be suggested that, on the evidence for effects of pollution, some other index would be better than these. They are used, not because it is assumed that they cannot be improved upon, but because they are reasonable given the available evidence and because they will demonstrate how an index can be used in the evaluation of various technological alternatives.

The effectiveness of a device depends upon the conditions under which the vehicle is driven. Emission rates vary widely for stop-and-go driving as compared with steady-speed driving, and between short hops and long trips. To compare cars on a uniform basis, California has established a procedure whereby the vehicle is placed on a dynamometer and run through a precisely defined set of accelerations, decelerations, and cruises. This test, intended to represent typical driving conditions, has been dubbed the California seven-mode cycle. Since it has been used for all federal measurements until and including 1971, all comparisons are on a uniform basis.

In 1970, the federal government adopted a new driving cycle and measurement technique, to become effective with the 1972 emissions standards. This constant volume sampling method, or CVS, produces numerically higher readings on the same vehicle than the seven-mode cycle. While an exact correspondence has not been established, it appears that cars which test 65 to 79 grams per mile of CO and 9 to 12 grams per mile of HC on the seven-mode test will produce 126 and 17 grams per mile of CO and HC, respectively, on a CVS test. Thus, for a given emission rate, the numbers will be 1.8 and 1.6 times higher.[3] After 1972, all emissions standards will be given in CVS readings.

The emission figures used here are intended to represent lifetime averages rather than new car data. In general, emissions seem to rise rapidly at first, then level off near 25,000 miles. Where averages over the vehicle life are not used, or are inappropriate, this will be noted.

Summary of Abatement Devices

Here, we will briefly describe twelve alternative abatement devices that may be adopted for pollution control. In the next section, these devices will be compared in terms of their relative cost and effectiveness.

POSITIVE CRANKCASE VENTILATION (PCV)

Crankcase vapors are not vented to the atmosphere, but are drawn into the carburetor and burned in the engine. The crankcase draft tube is replaced with an oil trap, hoses to the carburetor, and a PCV valve. Another hose connects the air filter with the breather cap. As a result, crankcase emission of hydrocarbons is completely eliminated.

1968 CLEAN AIR PACKAGE (68 CAP)

Changes are made in the settings of several carburetors and distributor parameters and in other engine design details. The apertures of the choke, idle fuel system, and main carburetor jets are all adjusted to a lean setting. The ignition timing is retarded at idle, and the idle speed increased. A hotter thermostat and a larger cooling system are specified. The shape of the combustion chamber and head gasket are altered, and the heating of incoming air is improved.[4]

CONTROLLED COMBUSTION SYSTEM (70 CCS)

In an extension of the 1968 CAP, the carburetor is again adjusted for leaner operation and is constructed to closer tolerances. Ignition timing is further modified and, in some models, is regulated in part by transmission operation. Crevices in the combustion chamber are reduced, and carburetor in-

3. J. N. Pattison, "Motor Vehicle Pollution Control News," *Journal of the Air Pollution Control Association,* 21, no. 2 (February 1971):89.
4. C. M. Heinen, "Vehicle Exhaust Control—Problems and Solutions," United States Senate, 90th Congress, Committee on Public Works, Hearings, *Air Pollution 1967 (Automotive Air Pollution),* pp. 425–451.

take air is preheated to a constant temperature on most models. Valve timing is somewhat altered.[5]

LOW-LEAD, LOW-OCTANE ENGINE (71 LLO)

The 70 CCS engine is modified to reduce oxides of nitrogen and to run well on low-lead gasoline. The compression ratio is reduced from the recent average of 9.5:1 to 8.5:1, with the result of reducing peak combustion temperatures and oxides of nitrogen. In addition, the spark advance is regulated in part by the transmission operation, so that oxides of nitrogen are further reduced. This requires a switch on the transmission and a solenoid in the vacuum advance line. In some engines, primarily large ones, the valve overlap is changed to increase the residual exhaust gas in the charge and again cut oxides of nitrogen.[6] Other changes are made in carburetors, intake manifolds, and the ignition system.

EVAPORATION EMISSION CONTROL (TANK)

The gasoline that normally evaporates in the gas tank and escapes through the vent is trapped, separated from the liquid, and sent to the carburetor to be burned. Gasoline that evaporates from the carburetor when a hot engine is shut off ("hot soak") is stored in a canister of activated charcoal and then drawn off when the engine is restarted. The equipment consists of tubing, separators, valves, and the charcoal canister.[7]

An alternative to this system is to modify the fuel composition to reduce its evaporation rate or its reactivity.

EXHAUST GAS RECIRCULATION (EXR)

A 1971 LLO engine is modified to burn entirely unleaded gasoline. A tube is added to bring exhaust from the tailpipe to the carburetor just above the throttle plate. No recirculation is used below 20 miles per hour cruise, on deceleration, or when the choke operates. At all other times, a constant 15 percent of the air entering the carburetor consists of exhaust gas. Some modifications are made in the ignition timing. Unleaded gas is used to prevent fouling of the carburetor and valves by exhaust gas. This is the

5. W. G. Agnew, "Future Emission-Controlled Spark-Ignition Engines and Their Fuels," Paper presented at the American Petroleum Institute, May 12, 1969, Chicago, Illinois. General Motors Research Report, GMR-880.

6. Joseph Geschelin, "71 Engines: Getting Them to Church," *Automotive Industries,* October 1, 1970, pp. 45–48.

7. Agnew, "Future Emission-Controlled Spark-Ignition Engines," pp. 6–8.

basis of the system that several manufacturers are using to meet the 1973 U.S. standards.

CATALYTIC EXHAUST CONVERTER WITH UNLEADED GAS (CAT)

A catalyst bed is installed in the exhaust system of an EXR vehicle. This may be a single bed in one location, or a double bed, perhaps with separate locations for the oxidizing and reducing catalysts, depending on the type of system employed. Secondary air injection also may be used. The carburetor is adjusted to produce raw exhaust gas with the proper carbon monoxide content for efficient functioning of the catalyst. Unleaded gasoline is used exclusively to avoid fouling of the catalysts. The catalysts must be replaced at specified intervals. This is the basis of the system that several manufacturers are projecting for compliance with 1975–1976 U.S. standards.

An alternative device for meeting these standards is the exhaust gas reactor, in which the catalyst is replaced by a large chamber, close to the engine, where hot exhaust gases are given time to complete combustion. While costs and performance may differ from the catalyst system, the catalyst now seems the preferred technology. In any event, the method of analysis for the two is essentially identical.[8]

GASEOUS FUEL CONVERSION (71 CNG)

A standard 1971 LLO vehicle is converted to burn compressed natural gas (CNG), as well as gasoline. Two 51-inch-long DOT 3AA2265 cylinders are installed in the trunk; an adapter is added to the carburetor; and various valves, regulators, and tubing are installed. A switch on the dashboard enables the driver to use either gasoline or CNG at any time. The CNG tanks must be filled from a special high-pressure source, since they will hold over 2,000 pounds per square inch. Trunk space is substantially reduced by the two tanks. Since the range on CNG is 80 to 100 miles, for intercity driving the gasoline tank undoubtedly would be used.[9]

Because of the limited range and special fueling needs, it is assumed that this conversion is used only in fleet automobiles such as taxis, government cars, and business fleets. These vehicles return to a central garage at

8. For a more complete discussion of catalysts and exhaust gas reactors, see Bain, "Automotive Air Pollution Control."

9. *Federal Low-Emission Vehicle Procurement Act,* Joint Hearings before the Subcommittee on Air and Water Pollution of the Committee on Public Works, U.S. Senate, 91st Congress, January 27–29, 1970.

night, and the requisite fueling equipment can be installed there at a modest cost per vehicle-mile served.

A number of fleets are currently powered by CNG, and it appears that with properly designed and installed fuel systems the danger of fire or explosion is no greater than with gasoline.

MODIFIED GASOLINE COMPOSITION (GCOMP)

The hydrocarbon content of gasoline can be modified, with some effect on vehicle emissions. Here, two alterations are evaluated and compared with a typical unmodified gasoline. In the first, the volatility of the fuel is reduced from 8.6 to 6.0 Reid Vapor Pressure, reducing evaporation. In the second, the lighter olefins through C_5 are replaced with saturated hydrocarbons. The engine used is a 1971 LLO without evaporative controls, but results would be little changed on older cars.

MINOR ENGINE TUNE-UP (TUNE)

A tune-up can cover a wide variety of adjustments from the simple and minor to complex operations requiring sophisticated equipment and highly trained mechanics. The tune-up considered here is a very minor one in which only engine idle parameters are affected. The idle speed is set to manufacturer's specifications, and the idle air-fuel ratio is properly adjusted. This operation is performed once a year and is considered an addition to whatever maintenance is usually performed.

ANNUAL POLLUTION MEASUREMENT WITH DIAGNOSIS AND CORRECTION OF HIGH EMITTERS (TESTUNE)

All vehicles are run once a year in an idle emissions test instead of the California seven-mode cycle. This simplified test gives emission rates for hydrocarbons and carbon monoxide, which can be interpreted to determine whether the vehicle can be improved by tune-up or repair. If it can be improved, the indicated action is taken and the vehicle retested. Standards are such that all properly tuned vehicles can pass the test, even though they may have no exhaust control devices installed. Results are not comparable with those of the seven-mode cycle.

USED CAR KIT (OLDKIT)

Makers of many cars offer a kit that includes some new parts, installation instructions, and special tune-up instructions for use on cars with no exhaust emission controls. The instructions require setting the idle speed at about 200 revolutions per minute above standard and adjusting the idle mixture leaner than the manufacturer's specifications. A sealer is provided to make it difficult to undo these adjustments. Ignition timing is set to the manufacturer's specifications and the vacuum advance disconnected, except that if the engine overheats it becomes operative automatically. Misfiring resulting from spark plug or wire failure is corrected. The kit can be applied to almost any pre-1968 car, or pre-1966 California car. It provides for older cars some of the features of the CCS vehicle.

Comparison of Abatement Alternatives

The basis for comparison of the devices considered here is a 1963 model car with no pollution controls (BASE). It is not a perfect standard because factors other than control devices—specifically, compression ratios—have since changed and affected emission rates. Still its emissions are probably representative of what present vehicles would produce if no controls were imposed.

The emission levels achieved by each of the twelve devices and the cost of those devices are calculated and presented in detail in Appendix C. Summaries of these results are given in Tables 7.2 and 7.3. Table 7.2 shows the emissions that result from each device or set of devices, by source, with totals for the vehicle. Where one device is used in addition to another, resulting in a cumulative effect, this is indicated in the last column. For example, system 6, exhaust recirculation, is applied to system 4, a vehicle that already includes the 1971 controls, and uses low-lead, low-octane gasoline.

Table 7.3 presents the added cost of each device and its marginal cost of abatement per degree of abatement, using the five indices of abatement developed in Chapter 3. The last column shows gross emissions expressed in Net Percent. Net Percent and Pollution Units reduced will be our basis for comparison of the twelve devices, with Pollution Units (PU) numerically equal to 0.01 times Net Percent.

In Table 7.4 the cost data of Table 7.3 are rearranged to show marginal and total cost for each device, and marginal and average cost per PU of abatement. Total cost of the devices is graphed in Figure 7.1, and marginal cost of all devices per PU is graphed in Figure 7.2. In addition to the costs generated in Appendix C, Table 7.4 and Figure 7.1 include 1971 cost estimates from the Environmental Protection Agency. The National Academy

Table 7.2. Emission Levels with Various Controls (grams/mile, seven-mode cycle)

System	Identification	Crankcase HC	Evaporation HC	Total					Done to System No.
				Exhaust				HC	
				HC	CO	NO$_x$	Lead		
0 Uncontrolled Car	BASE	3.15	2.77	10.20	76.9	4.0	0.133	16.12	
1 Positive Crankcase Ventilation	PCV	0	2.77	10.20	76.9	4.0	0.133	12.97	0
2 1968 Clean Air Package	68 CAP	0	2.77	4.38	38.3	6.08	0.133	7.15	1
3 1970 Controlled Combustion System	70 CCS	0	2.77	3.45	25.0	7.0	0.133	6.22	1
4 1971 Low-Lead, Low-Octane	71 LLO	0	2.77	3.45	25.0	4.5	0.027	6.22	1
5 Evaporation Emission Control	TANK	0	0.49	——— same ———					any [4]
6 Exhaust Recirculation	EXR	0	2.77	3.45	25.0	2.07	0	6.22	4
7 Catalyst, No Lead	71 CAT	0	2.77	0.69	2.3	2.07	0	3.46	6
8 Natural Gas Fuel Conversion	71 CNG	0	2.77	0.6	7.0	0.6	0	3.37	4
9 Alter Gasoline Comp	GCOMP	0	1.11	——— same ———					any [4]
10 Minor Tune-up	TUNE	0	2.77	3.95	32.2	6.44	0.133	6.72	1,2
11 Test, Diagnose, Tune-up	TESTUNE	0	2.77	— depends on model year —					1,2
12 Used-Car Kit	OLDKIT	0	2.77	5.1	53.8	2.8	0.133	7.87	1

Source: Appendix C.

Table 7.3. Marginal Costs of Various Controls

System No. Ident.	Compared with	Cost Capital ($/year)	Variable ($/mile)	Total ($/mile)	2 $ Net %	3 $ min max %	4 $ % Reduced	5 $ Unleaded Net %	6 $ Unleaded min max %	Gross Emissions (Net Percent)
1 PCV	0	0.680	0.00020	0.00027	0.54	∞ a	0.130	0.40	∞ a	95.0
2 68 CAP	1	2.176	0	0.00022	0.25	∞	0.024	0.19	∞	86.5
3 70 CCS	1	3.4	0	0.00034	0.40	∞	0.030	0.30	∞	86.5
4 71 LLO	3	4.00	0.0010	0.00140	0.39	∞	0.098	0.67	∞	51.0
5 TANK	4	4.08	0.00020	0.00061	1.74	∞	0.44	1.30	∞	47.4
6 EXR	4	9.65	0.0011	0.00206	1.01	∞	0.25	1.01	∞	30.7
7 71 CAT	6	26.93	0.00239	0.00508	3.06	∞	0.76	3.08	∞	15.0
8 71 CNG	4	55.77	(−0.0064)	(−0.0008)	all negative					11.3
9 GCOMP	4	0	0.00074	0.00074	2.86	∞	0.72	2.15	∞	48.0
10 TUNE	1,2	0	0.00030	0.00030	7.50	∞	0.28	5.66	∞	86.1
11 TESTUNE	1,2	0	0.00247	0.00247	3.01	∞	0.75	2.25	∞	85.0
12 OLDKIT	1	8.31	0.00036	0.00119	0.52	∞	0.13	0.39	0.40	69.5

Source: Appendix C.
a. Emissions are not reduced in most cases where measured by indices 3 and 6, so with a positive cost the marginal cost per percentage reduction is infinite.

Table 7.4. Abatement Cost per Pollution Unit

System No.	Marginal Cost ($/mile)		Total Cost ($/mile)		Gross Emissions (units)	Cost per Unit	
	Own[a]	EPA[b]	Own[a]	EPA[b]		Marginal ($/unit)	Average ($/unit)
1	0.00027	0	0.00027	0	0.950	54	54
2 (+1)	0.00022	0	0.00049	0	0.865	24	36
3 (+1)	0.00034	0	0.00061	0	0.865	40	45
4 (+3+1)	0.00140	0	0.00201	0	0.510	39	41
5 (+4+3+1)	0.00061	0	0.00262	0	0.474	174	50
6 (+5+4+3+1)	0.00206	0	0.00468	0.00136	0.280	101	65
7 (+6+5+4+3+1)	0.00508	0.00508	0.00976	0.00671	0.150	306	115
8 (+5+4+3+1)	(—0.0008)	—	0.00388	—	0.090	(—)	43
9 (+4+3+1)	0.00074	—	0.00275	—	0.480	286	53
10	0.00030	—	—	—	0.861	750	—
11	0.00247	—	—	—	0.850	301	—
12	0.00036	—	—	—	0.695	52	—

a. Costs from Table 7.3.
b. Costs from Environmental Protection Agency, *The Economics of Clean Air*, Report of the Administration to Congress, Senate Doc. no. 92-6 (Washington, D.C.: U.S. Government Printing Office, March 1971), p. 3–16.

of Sciences provides still another set of cost-effectiveness data. They estimated the abatement demanded by each set of standards from 1968 to 1976 and estimated the cost of the devices used and proposed to meet these standards. See Table 7.5. While both costs and abatement are somewhat different from those of Table 7.4, the same trends appear. Marginal and total costs rise rapidly after 50 percent abatement, the level achieved in 1971 or 1972.

Figure 7.1 shows clearly the rapid rise in cost that accompanies higher levels of pollution control. The cost rises, not linearly, but at an ever-increasing rate, especially after 50 percent control. If the curve were extrapolated beyond the last data point, it might well be asymptotic to the 100 percent abatement line, demonstrating the impossibility of building a perfectly clean car. The Environmental Protection Agency costs, while lower than those computed here, rise just as rapidly after 1971. They show no cost until 1971 because they assume that with current experience in design-

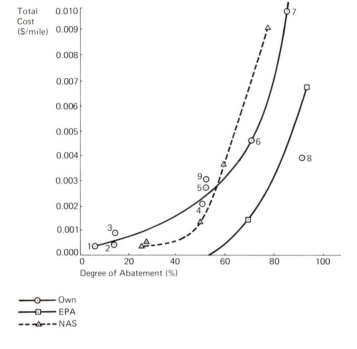

Figure 7.1. Total Abatement Cost of New Car Devices
Source: Tables 7.4, 7.5.

ing and manufacturing control systems, the devices up to that time could be included at such a small increase in cost that the fuel savings from lean operation would more than offset them. In addition, the EPA does not include as a cost the loss of power caused by 1971 and later controls. The National Academy of Sciences curve, using the most recent data, rises most steeply of all.

The marginal cost curves of Figure 7.2 permit a comparison of the cost effectiveness of the alternative devices. The line drawn through some of the new car device data points shows again the relative constancy, or random variation of costs through 1971, and the sharp increase after that. The marginal cost per PU of system 6, EXR, is over twice that of the 1971 LLO vehicle, and the marginal cost per Unit of the catalyst, system 7, is three times that of EXR. For both these devices, the costs are about equally di-

Table 7.5. Abatement Cost per Pollution Unit (Data from NAS)

Model Year	Marginal Cost ($/mile)	Total Cost ($/mile)	Gross Emissions (units)	Cost per Pollution Unit	
				Marginal ($/unit)	Average ($/unit)
1968	0.00025	0.00025	0.75	10.0	10.0
1970	0.00011	0.00035	0.72	36.7	12.5
1972	0.00094	0.00129	0.50	72.3	25.8
1973	0.00236	0.00366	0.42	295.0	63.1
1975	0.00537	0.00902	0.27	358.0	123.6
1976	na	na	0.05		

Source: Cost and emission data from the Semiannual Report of the Committee on Motor Vehicle Emissions of the National Research Council to the Environmental Protection Agency, National Academy of Sciences, Washington, D.C., January 1972. Fuel consumption is assumed to increase 3 percent in 1971, 6 percent in 1973, and another 9 percent in 1975. Lead is not considered, so units are equal to unleaded net percent divided by 100.

vided between increases in production cost and increases in cost of maintenance and operation, as shown in Table 7.3. The NAS data show even more pronounced marginal cost increases after 1970.

Of the two devices that reduce evaporative emissions, system 9, GCOMP, is clearly dominated by system 5, TANK. The installation of evaporative controls thus seems to be much more cost-effective in reducing hydrocarbons than is a change in gasoline composition. Using both together is more effective than either alone; but since the marginal cost of the combination would be five times higher than that shown here, on a per PU basis, it is most unlikely that it would be justified except in the most polluted areas. Both have high costs per PU as compared with all other new car programs; but this is probably overstated for TANK, which is reported to be less expensive now than the original costs on which this analysis is based, as reflected in the NAS curve, below TANK.

A most interesting set of results is that for the CNG automobile. Compared with the 71 LLO, this vehicle is actually cheaper overall, despite the large capital investment, and it has the lowest emissions of any system reported on. It must be remembered, however, that because its fuel tanks are heavy and bulky and because it needs special fueling stations, it is practicable only for a limited set of situations, primarily for fleet vehicles that return daily to a central garage for fueling. Also, the cost computation is for

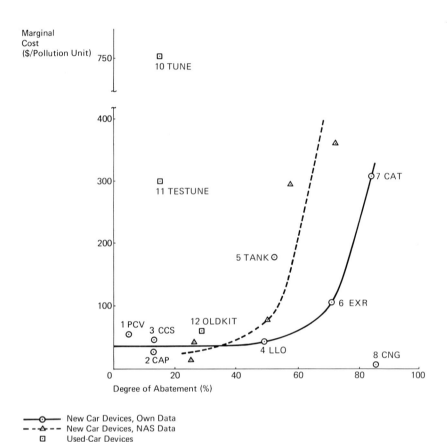

Figure 7.2. Marginal Abatement Cost per Unit
Source: Tables 7.3, 7.4, and 7.5.

recent natural gas consumption rates at 1971 prices rather than the ex-
panded demand that would result from widespread automotive conversion.
It does not reflect any shift in the demand for natural gas for other uses
resulting from environmental demands. Thus it is not recommended, or
even suggested here, that the bulk of U.S. automobiles be built for CNG
operation.

It does seem that in large cities, where high pollution levels and residential densities mean large benefits from a given unit of abatement, it would

be worthwhile to encourage use of CNG by all for whom it is practicable. This could include taxis at a minimum, and most commercial and private fleets at a maximum.

Of the systems applicable to used cars—numbers 10, 11, and 12—OLD-KIT is close to early new car systems in cost/percent, while TESTUNE is as costly as 1975 controls, and TUNE is far more expensive. The reason for this seems to be that TUNE merely sets engine idle parameters to the manufacturer's specifications while TESTUNE repairs ignition and carburetor malfunctions. The percentage reductions in hydrocarbons and carbon monoxide achieved in this way can be significant, but not enduring, since the adjustments or repairs may have to be repeated annually.

This suggests that, while improper maintenance may account for a significant proportion of the emissions of used cars generally, the identification and repair of "high emitters" is expensive considering the longevity of its effect.

OLDKIT differs in that it reduces emissions, even for old automobiles in good repair, by setting new specifications for several easily adjusted engine parameters. The resulting emission reductions are quite large, and the operation, while expensive, need be performed only once rather than annually. On automobiles more than eight years old, OLDKIT will be expensive, since its cost must be spread over only a few years; but for those in the five- to eight-year-old category, it is clearly preferable to the annual tune-ups.

If Net Percent is the appropriate measure of abatement, total automotive emissions have been reduced by about 50 percent by the 1971 vehicles. The EXR system, which approximates the 1973 U.S. standards, brings another significant reduction in emissions primarily by reducing lead and oxides of nitrogen. Total emissions drop to about 30 percent of BASE, but the total cost doubles as a result. The marginal cost per PU is double that of all previous systems except for TANK (see Figure 7.2). The installation of the CAT system, approximating 1975 U.S. standards, again doubles the total cost of abatement, and the marginal cost per PU is about triple that of EXR.

If the degree of pollution control is to be chosen so that the marginal cost of abatement is equal to the marginal benefits, then clearly, as marginal abatement costs rise, the marginal benefits necessary to justify that

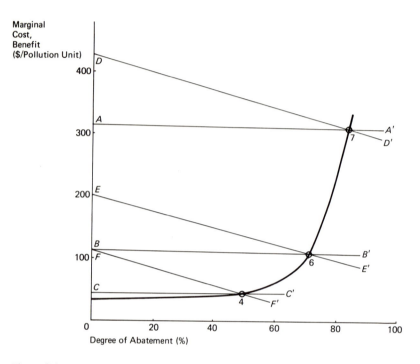

Figure 7.3. Comparison of Marginal Costs and Benefits

degree of control also rise. Figure 7.3 shows the marginal abatement cost curve derived in Figure 7.2. The horizontal line *AA'* shows the marginal benefits necessary to justify use of the CAT system, assuming that benefits are a linear function of emission rates, so that marginal benefits are constant. Line *BB'* shows the benefits required to justify the EXR system, and *CC'* the benefits for the 1971 LLO vehicle. Distance *BA* is the increase in marginal benefits necessary to justify a move from EXR to CAT, in this case $205.

In Chapter 3, we concluded that benefits, in general, rise more rapidly than emissions, if other things are held constant; so marginal benefits will decline as emissions decline. The downward-sloping lines in Figure 7.3 represent marginal benefits that decrease with abatement, implying a non-linear total benefit curve. These are drawn to intersect points 4, 6, and 7, so that they represent benefit levels to justify the three control devices under consideration, although the degree of slope is arbitrary. Because the emissions of system 6, EXR, are greater than those of 7, CAT, the former is left of the latter, and distance ED is greater than distance BA. This implies that at a given pollution level, a greater difference in marginal benefits is required to justify CAT over EXR than in the linear benefit case, since further abatement reduces marginal benefits. The same relationship holds for the difference between line EE' and FF', although here the difference is greater because the horizontal difference between points 4 and 6 is greater than that between points 6 and 7. Rising marginal costs and falling marginal benefits imply that great differences must be found in marginal benefits to justify large increases in marginal abatement cost.

The effect of nonlinear benefits on choice of degree of abatement can best be seen by looking at a time trend of emissions under different abatement assumptions. Figure 7.4 shows relative annual emissions of the four pollutants under different abatement schedules. Curve a assumes that all standards through the original 1975–1976 standards are applied on schedule. Curve b shows what would happen if all standards were applied, but the 1975–1976 standards were delayed to 1985 and 1986. Curve c applies to the original standards through 1973, the interim standards in 1975, and the original 1975 and 1976 standards in 1985 and 1986.

Under plan a, all emissions decline until about 1985. Since most emissions under plan a will be at less than one-half their current level by 1981, and three will be below 20 percent; the marginal benefits of abatement, given a nonlinear function, will be much less than they are now. In setting abatement levels, then, it is necessary to consider not only current pollution levels and benefits, but levels that will prevail five or even ten years from now. If marginal benefits of abatement are now, and will be, at pollution levels 80 percent lower, over \$300 per Pollution Unit, it would seem desirable to proceed with plan a, provided that costs and performance are assumed here and that marginal costs of 1976 standards are not greater than those for 1975—a highly unlikely assumption. But if benefits may drop well below this level during plan a, some consideration should be given to a less costly alternative, perhaps plan b or c.

Suppose, for example, that current marginal benefits were \$200 per Pollution Unit and were proportional to emissions. Even by 1975, when the new standards should come into effect, marginal benefits would be less

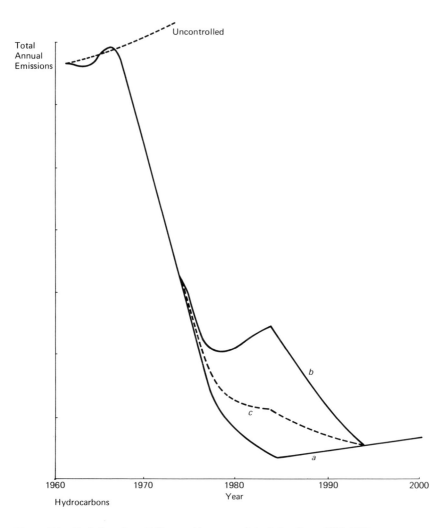

Figure 7.4a. Emissions from Different Abatement Schedules, Years 1960–2000

Note: These curves assume that emission standards are imposed on the following schedule:
a. All standards applied as originally scheduled through 1976.
b. As in *a*, except that the 1975 and 1976 standards are postponed to 1985 and 1986.
c. All standards applied as originally scheduled through 1973, interim 1975 standards
   applied in 1975, original 1975 and 1976 standards applied in 1985 and 1986.
The emission rates for each standard are taken from the Semiannual Report of the
Committee on Motor Vehicles Emissions of the National Research Council to the
Environmental Protection Agency, National Academy of Sciences, Washington, D.C.,
January 1972, except that uncontrolled emissions of nitrogen oxides are assumed to be
4 grams per mile in 1960, 6 grams per mile in 1970, and 4 grams per mile in 1971. It is
assumed that annual new car sales increase by 4 percent each year.

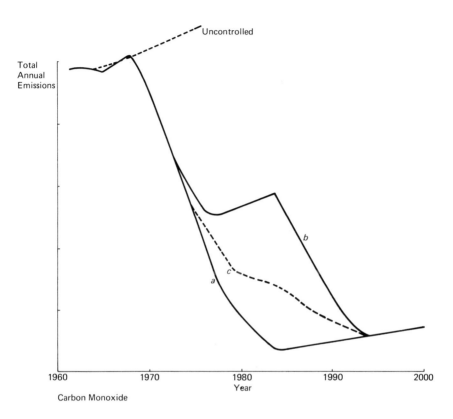

Figure 7.4b. Emissions from Different Abatement Schedules, Years 1960–2000

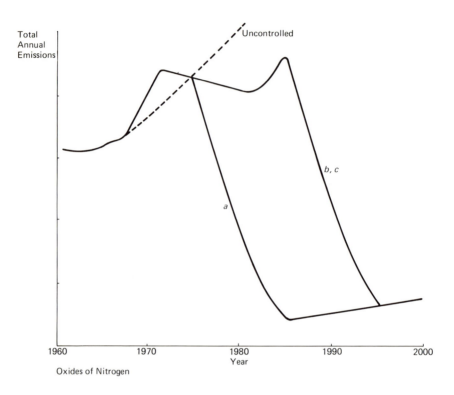

Figure 7.4c. Emissions from Different Abatement Schedules, Years 1960–2000

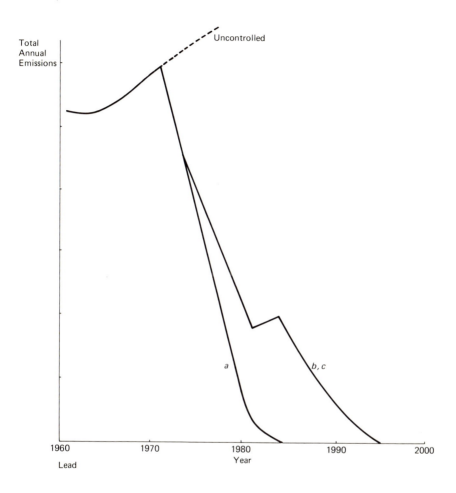

Figure 7.4d. Emissions from Different Abatement Schedules, Years 1960–2000

than half that, or $100, not justifying CAT and hardly justifying EXR. Imposing the 1973 standards would keep marginal benefits in the same range until the mid-1980s, when they would begin to rise again. This would suggest that the optimal policy would be to follow plan b, deferring the 1975–1976 standards until 1986. This argument is an example only, since marginal benefit figures are not available; but it suggests the impact of long-term changes in pollution levels on the choice and timing of policies.

The importance of the above timing issue can best be seen by looking at aggregate costs in place of marginal figures. Suppose that it were felt that for one-half of the United States the 1975 standards were not necessary until 1985, so that half of all cars could be built to 1973 standards for an extra decade, following plan b. If an average production of 10 million cars per year is assumed, 50 million vehicles would be exempted at a savings of $0.00508 per mile[10] for a lifetime of 100,000 miles, or $25.4 billion at current prices. With potential savings of this magnitude, it is absolutely essential that alternatives such as a delay in standards or a two-car policy be considered. Even plan c, with its lower emission rates, would allow some savings over plan a.

Unconventional Propulsion Systems
All of the pollution controls considered above involve modifications to the gasoline-fueled internal combustion engine (ICE). In recent years, however, increasing attention has been given to the development of other propulsion systems that may differ radically from the ICE and may provide much lower emission rates. Space does not permit an exhaustive study of the non-ICE possibilities here, but the most promising contenders can be briefly reviewed.

The Wankel engine is an ICE that uses a triangular rotor in place of pistons to compress the air-fuel mixture and translate the energy of combustion into rotational energy. Some imported vehicles currently use the Wankel engine, and General Motors is investigating its use for domestic cars. It has been suggested that, while the Wankel produces more hydrocarbons and carbon monoxide than conventional engines, it is easier to remove these from its exhaust, so that it will have less difficulty meeting the 1975

10. See Table 7.4. Actual savings would be greater because of the 1976 standards, not costed here.

and 1976 standards than will conventional engines. Fuel consumption appears to be greater than for comparable piston engines. While it is too early to estimate costs of using the Wankel to replace piston engines and meet future pollution control standards, this is one new development already in use that may grow rapidly in the near future.

The diesel engine is a high-compression cousin of the gasoline ICE in which ignition is provided, not by a spark, but by heat generated in the compression stroke. It has been in existence since the turn of the century and is used widely in buses and trucks. Diesels produce less HC and CO than a gasoline engine, but as much or more NO_x. In addition, diesel exhaust has a more objectionable smell than that of the gasoline engine, so that it is not clearly preferable from an environmental viewpoint. The diesel engine is more expensive and heavier per horsepower than a gasoline engine but uses less fuel and is the most economical choice for heavy-duty applications.

The gas turbine is essentially a jet turbine in which the exhaust, instead of providing thrust directly, spins turbine blades that can drive the vehicle through gear reductions. Large gas turbines produce high power-to-weight ratios—a feature that has made them attractive for use in aircraft—and they also require much less maintenance than the piston engines that preceded them. When reduced to the size appropriate for automobiles, however, gas turbines become less efficient and produce more pollution than in aircraft applications. HC and CO are lower than for the ICE, but NO_x may be higher. Chrysler has been building turbine-powered prototype automobiles for years, but their performance and high cost render them impractical for the present.[11]

The Rankine vapor-cycle external combustion engine, or steam engine, continues to be attractive to the public and to engineers. Quiet operation, inherently low pollution emissions, and smooth power production without a transmission are all regarded as important advantages for this engine. There are numerous technical problems, however, including the need for a separate engine to run auxiliaries (such as the generator, power steering, air conditioning), slow warm-up, possible freezing in winter, lubrication, and the complicated valving needed to achieve both high power and high

11. Robert U. Ayres and Richard P. McKenna, *Alternatives to the Internal Combustion Engine* (Baltimore, Md.: Johns Hopkins Press, 1972), p. 255.

efficiency. At the present time, no steam engine has been demonstrated that will give the long life of an internal combustion engine under actual automobile operating conditions at comparable cost. Recent steam power plants are both heavier and bulkier than their internal combustion counterparts. The steam engine is thus not a practical alternative for the present.

The electric vehicle, using an electric motor powered by storage batteries, also has captured the imagination of the public and of some researchers. Its quiet, smooth operation and lack of visible pollution (since the electric generation station is remote) make it seem the ideal vehicle. But the electric vehicles that can be built under current technology have low power, low top speed, short cruising range, and are generally more expensive than gasoline-powered counterparts. Some are referred to as specifically urban vehicles because of these performance limitations. The implicit cost of replacing the ICE with a battery-electric system would be enormous and would certainly be dominated by the CNG-powered car at present technology. At least in the short run, batteries will power only a small percentage of all vehicles, and these will be for very special applications. There is no suggestion that an electric car can be produced before 1980 that will be able to match most of the functions now performed by automobiles.[12]

Attempts have been made to capture the advantages of the electric vehicle and the ICE by building hybrids in which a small ICE runs continuously at steady speed, powering a generator that supplies propulsion power and charges batteries when the drive motor does not demand all its power. When high power is required for acceleration, the batteries are discharged. This combination can be much cleaner than the standard ICE vehicle; the highest emissions come from changes in engine power and speed, but in this case both are provided by the batteries. The hybrids have a much greater cruising range than battery electrics, but they are much more expensive and heavier than their ICE counterparts. It is unlikely that a hybrid electric can compete with a standard ICE in either performance or initial cost, so it is not a practical alternative for the present.[13]

12. *The Automobile and Air Pollution: A Program for Progress*, Report of the Panel on Electrically Powered Vehicles (The Morse Report), (Washington, D.C.: U.S. Department of Commerce, 1967), pp. 67, 88; and U.S. Department of Health, Education and Welfare, Public Health Service, *Control Techniques for Carbon Monoxide, Nitrogen Oxide, and Hydrocarbon Emissions from Mobile Sources* (Washington, D.C.: U.S. Government Printing Office, 1970), p. 7–14.
13. Ayres and McKenna, *Alternatives to Internal Combustion Engine*, p. 237.

Some estimates have been made of the specifications, cost, and emission rates for alternatives to the gasoline engine and these are presented in Table 7.6. The figures must be considered with caution. The specifications are estimated on present technology, but on the assumption that a number of technical problems can be worked out. For example, the steam engine efficiency, weight, and cost data are based on the kind of system that is currently built; but it is assumed that the problems such as lubrication, auxiliary power, selection of a working fluid, and others can be solved at a small cost. In addition, the performance and durability of all vehicles are not comparable. The range and top speed of the electric car are greatly restricted compared to those of the alternatives;[14] and the acceleration rate and noise level of the diesel are worse than those of the gasoline engine. This table therefore cannot be used to select the best solution to the automotive pollution problem. At best, it indicates some characteristics that may be achieved if many current problems are solved, and it ignores performance entirely. It does not alter the conclusion that there is no immediate substitute for the gasoline-powered ICE, for most automotive uses.

It seems possible that during the 1970s some of the technical problems of these power plants will be solved so that a gas turbine, steam engine, electric drive, or hybrid vehicle will appear as an attractive alternative to the ICE vehicle, for at least some purposes. If and when it does, it can be compared to other emission control alternatives. At a time when the specification and performance of these vehicles are significantly worse than those for conventional power plants, there is either no basis for comparison, or the implicit costs are very high—certainly above costs of some low-emission internal combustion engine such as the catalytic reactor or CNG engine.

Even if a prototype of a turbine, steam, electric, or hybrid vehicle were revealed today with performance and costs comparable to those of an ICE and with lower emissions, it would still be years before it could account for a significant portion of new car sales. History shows that laboratory prototypes tend to run well only with extensive tinkering and adjustment —in other words, with high maintenance costs. Frequently they have behav-

14. One manufacturer currently offers a battery-electric using a conventional subcompact chassis. The battery version costs about $1,500 more than the ICE model and has a top speed of 70 MPH and a range of 60 miles. Even this limited range is attainable only at moderate speeds, however, and at 70 MPH it would certainly be much less. (*The Toronto Star*, October 23, 1972, p. 13.)

Table 7.6. Comparison of Alternative Propulsion Systems: Specifications, Cost, Emissions

Engine	Max. Energy Conversion Efficiency[a] (%)	Total Weight Power Train (lb)	10-Year Cost Change[b] ($)	Pollution Emission Rates (grams per mile)				
				HC	CO	NO$_x$	SO$_x$	Particulates
Gasoline (before controls)	19	750	0	11.5	85	4	0.27	0.36
Diesel	21	900	−902	3.5	5	4		
Gas Turbine	17	400	+830	0.2–0.9	2–8	1.0–1.6	0.3	
Steam Engine	14	715	+180	0.13–0.4	0.35–2.0	0.25–0.6	0.2–0.3	
Electric	30	1,500	+912	negl.[c]	negl.[c]	1.75[c]	0.002[c]	0.07[c]
Hybrid (gasoline-electric)	17	1,275	+774	0.4	4	0.4	—	—

Source: Robert U. Ayres and Richard P. McKenna, *Alternatives to the Internal Combustion Engine* (Baltimore, Md.: Johns Hopkins University Press, 1972), Chap. 14, Tables 14.2, 14.3, 14.8.

Note: These figures assume that a number of technical problems which currently exist can be solved. In addition, the performance of the vehicles varies widely, with some offering limited top speed, limited range, limited acceleration, or all three. Thus, even if these specifications were met in production vehicles, they would not necessarily substitute for the ICE in a wide range of uses.

a. This is the percentage of the energy content of the fuel that becomes available at the rear wheels for propulsion in a typical driving cycle. The figure for electric vehicles is the thermal efficiency of the central generating station and does not include distribution losses or the losses in charging and discharging the batteries.

b. The sum of capital cost differences plus changes in operating and maintenance costs over a ten-year vehicle life.

c. Assuming that the electric generating station burns natural gas, by far the cleanest fuel but not in widespread use in such plants.

ioral quirks that only the inventor or a select portion of the scientific and engineering community can cope with. The development of such an engine into one that will operate reliably for years with infrequent maintenance, performed by mechanics of average skill levels, and requiring little special understanding on the part of the driver, usually takes years.

The larger and more complex the new product or component and the greater its differences from past counterparts, the longer the process takes. This suggests that radically new power plants are not likely to be widely used any sooner than five or ten years after the testing of a feasible prototype. The ICE, whether piston or rotary, will power most automobiles at least to the end of this decade.

Conclusions

It must be remembered that the systems investigated in this chapter are not all of those that will be technically feasible in the next few years but are a sampling designed to indicate the general limits of current and emerging technology. It is possible that some devices will emerge, or already exist, that can achieve at less cost the same abatement as exhaust recirculation or catalytic reactors. It is also possible that the latter two systems will not perform as well on production vehicles as tests and data indicate, or that their cost will be much higher or lower than we anticipate. Our analysis exemplifies how such technology should be evaluated and is not intended to prove the desirability or undesirability of a particular device.

Subject to these limitations, we have shown the cost effectiveness of emission reduction by new car devices, retrofit of used cars, and used-car maintenance. It is clear that at any point in time, marginal abatement costs rise with increased abatement and that this rise is precipitous approaching the abatement originally specified for 1975. The marginal cost seems to shift downward over time, because of technological development and learning by doing in production; but it is difficult to predict how rapidly such shifts will take place. Despite cost reductions, the marginal cost curve remains steep near the limits of technological feasibility. As abatement increases and costs rise, selecting the proper degree of abatement becomes increasingly important.

Modifications in new cars can probably bring overall emissions to 15–25 percent of base levels, at a cost in the vicinity of $100 per Pollution Unit.

Any further reduction will be more expensive per Pollution Unit, and meeting the original 1975–1976 standards appears difficult at any cost.

Radical changes in fuel, such as conversion to CNG, can cut emissions to less than 10 percent of base levels, at little extra cost for the special fleets that can conveniently use this fuel.

Modification of used cars can cut emissions by about 15 percent at a cost of about $40 per Pollution Unit, while tune-ups of used cars, under some conditions, cost about as much as projected 1975 controls. It does not seem possible to achieve large pollution reductions in used cars at any lower cost than that of new car programs, although this may change when more fragile catalyst systems are introduced.

There are thus three technological paths upon which abatement programs may proceed. In Chapter 8, we will consider these and other solutions and the administrative means whereby they may be implemented at local, state, and national levels.

Evaluation of Feasible Abatement Strategies

8

Scope of Evaluation

In the preceding chapters, we have explored the technological methods available for reducing automotive air pollution. Ideally, the next step would be to solve a mathematical model of the automotive pollution production function to find the optimum degree of abatement and the technology that should be used to achieve it. Then we should examine the administrative alternatives to find those instruments that could best achieve the desired optimum point.

In fact, the technological alternatives examined in Chapters 5 to 7 do not lend themselves to simple mathematical formulation, nor can the administrative alternatives be easily described. Rather than abstract the actual problem to a workable but unrealistic mathematical model, we will have to use less rigorous methods in evaluating abatement strategies.

The theory of externalities suggests that the optimum pollution abatement program is to select an emission rate such that the marginal cost of abatement is just equal to the marginal benefits of abatement.[1] In Chapter 3, we found that a benefit function cannot be precisely estimated. We can, however, suggest a wide range of possible health benefits from automobile pollution control ($1.5 to $8 billion per year) and indicate the shape of the marginal benefit curve, which rises with increasing pollution density and population density. Furthermore, the abatement costs discussed in Chapters 5 to 7 describe present technology; and these may change at any time as new technology becomes available. There is little hope, therefore, that a degree of abatement can be found that precisely equates these margins.

While it is not now, and may never be, possible to determine analytically the precise amount of automobile pollution that should be allowed, many other parameters of a pollution control program can be determined from this study. These parameters include the relative degree of control in various parts of the country, the technological abatement avenues to be followed, and the administrative devices that will best achieve the desired control at the least cost. The discussion that follows is addressed, not to specifying proper standards, but rather to finding the administrative means that will lead to the best balance of costs and benefits throughout the coun-

---

1. Allen V. Kneese and Blair T. Bower, "Causing Offsite Costs to Be Reflected in Waste Disposal Decisions," in Robert Dorfman and Nancy S. Dorfman, *Economics of the Environment* (New York: W. W. Norton and Co., Inc., 1972), pp. 135–154.

try over a period of time, given uncertain benefits and rapidly changing technology. This question may be more difficult to answer, but in the long run its answer is far more useful.

Composition of Total Strategy

If one were to determine a national average of abatement benefits, the value of a unit of abatement would be higher in highly polluted or densely populated areas, and particularly high in areas that are both. Sparsely populated and clean areas will have significantly lower benefits. Thus an optimal abatement program must require a higher degree of abatement in dense and dirty areas than in sparse and clean ones, if marginal costs of abatement are to be related to marginal benefits. Because the severity of air pollution depends not only on total emissions but on meteorological conditions such as wind velocity and sunlight, the correlation between population density and pollution density will be far from perfect, and they must be identified separately. Finally, conditions such as temperature may influence the effectiveness of some control devices and thus their marginal benefit. Devices designed for operation under warm conditions when photochemical smog is most severe may permit much higher emissions in cold winter weather and thus may increase the marginal cost of abatement in cold climates.

In Chapter 7, we found that the greatest cost effectiveness in pollution control is currently in the design of new cars, although these costs are rising as abatement increases. Maintenance of used cars seems to be less cost-effective than current new car programs. General restrictions on use of vehicles are an expensive means of reducing pollution, except in small areas such as individual streets or when reduced congestion is a valuable by-product. Finally, it appears feasible now to produce automobiles to more than one standard of cleanliness, although the total number of standards that can be met at a reasonable cost is small.

The variation in marginal benefit from one area to another and the variety of technological alternatives and their costs suggest that an optimal abatement strategy must consist of more than a single national new car standard. One standard was satisfactory when the cost of abatement was relatively low; but now that costs exceed $50 per year per car or $4 billion per year (for 1973 standards; 1975 may cost twice as much), it is clearly necessary to consider ways to avoid waste in the program. The attack on automobile pollution must consist of several components.

To devise these programs, the country can be divided into three kinds of areas, depending on marginal benefits of abatement: low benefit, medium benefit, and high benefit. Low-benefit areas are rural areas and small towns where population density is low and automobile pollution never reaches annoying or even perceptible levels. This category accounts for most of the geographical area of the United States, but only a small percentage of the population and the automobiles.

Medium-benefit areas are those where automobile pollution is a notice-able problem, but not a serious one. This includes most major metropolitan areas where population density is substantial, but pollution is moderated by adequate ventilation, a cool climate, or high transit usage.

High-benefit areas are those where automobile pollution is a serious and acute problem because of moderate or high population density and high pollution levels. These are the areas where photochemical smog is most prevalent, such as the Los Angeles basin. Perhaps a dozen major metro-politan areas would fall into the high-benefit definition, accounting for 20 to 25 percent of the population.

Since the federal new car program will be the major force for techno-logical progress in automobile pollution control, it should be aimed, not at the low-benefit, but at the medium-benefit areas; and the marginal abate-ment cost should approximate marginal benefits in these areas. Low-benefit areas include a small percentage of all vehicles, and by definition they are widely scattered. Since marginal abatement costs are low and relatively constant up to 50 percent abatement or more, as shown in Figure 7.2, the savings from establishing a separate program for low-benefit areas would be relatively small. The administrative problems would probably not jus-tify these small cost savings.

All automobiles sold in the United States should be required to meet at least the national program based on the medium-benefit areas. This na-tional new car program should include incentives to design the vehicle so that it has low emissions not just when it is new, but when it is old, given the maintenance provided by the average motorist and the average me-

chanic. That is, the marginal benefit should be based on the best estimate
of expected lifetime emissions.

Since the most cost-effective means of pollution reduction is modification
of new car design, even at high abatement levels, the high-benefit states
and regions should have their own new car programs. A single policy may
be set by all of them which is designed to equate the marginal cost of abatement with the high marginal benefit. In fact, to avoid proliferation of
regulations, the federal government may set two: a medium policy, which
must be met by all automobiles; and a stringent policy, which must be met
by automobiles to be sold in states or areas that specifically adopt it. The
state or area thus has a single choice: Does it adopt the stringent policy or
accept the medium?

Low-benefit areas, presumably, will take no independent action for pollution control, since their benefits are probably below the marginal cost of
any of the technical possibilities beyond the most rudimentary new car
devices. In the medium-benefit range, however, there are several technological possibilities in addition to new car design. These areas may consider
encouraging or requiring the retrofit of devices on used cars, or on those
cars where the cost per Pollution Unit is predictably highest. They may
also encourage the use of gaseous fuels by those fleet vehicles for which
they are appropriate. Finally, moderate programs of motoring reduction
may be undertaken.

High-benefit areas have a number of alternatives in addition to new car
design to bring the marginal cost of abatement up to marginal benefits
along all avenues of control. They should establish programs to ensure
retrofit of used cars, and perhaps maintenance of selected used cars, choosing carefully an alternative that has high cost effectiveness.[2] Encouragement
of gaseous fuels for fleet vehicles is essential, and there may be a payoff in
regulation of fuel composition, at least until most cars have evaporative
control systems. If catalysts are used in new cars, prohibition of leaded gasolines will be mandatory. An effective program to reduce total motoring
also may be employed, aimed at both the short and the long run. Emergency
strategies may be designed to further reduce the use of cars on days when
pollution reaches danger levels.

2. See Chapter 7.

Evaluation of Administrative Devices

The technological approaches to pollution control that may be pursued in different situations have been discussed above; the best policy instrument to implement the desired change must now be determined, by using the criteria of Chapter 2 as a guide. The selection will be made among the seven instruments listed in Chapter 2; the few for which there are no technological alternatives or that are not distinguished in important ways from others will be omitted.

NATIONAL NEW CAR PROGRAM

Pollution control agencies can choose between two fundamentally different strategies for regulating the emissions of new cars. The first, which has been pursued thus far by the federal government and the state of California, is the use of a regulation or constraint on the manufacturer, vendor, or operator, specifying the maximum amount of pollution that may be emitted by a motor vehicle. The second is a charge in the form of a tax or fine, relating to the emission characteristics of the vehicle. This may be imposed at the time of sale, annually, or on a mileage basis. In theory, the same degree of abatement can be achieved by either of these methods; but, in practice, their operation may be quite different. In this section, present practice under the constraint with the operation of a hypothetical effluent charge will be compared.

PRESENT CONTROL PROCEDURES In general, emission standards such as those used by California and the federal government are not set solely by reference to air quality objectives. Instead, they reflect a judgment as to the degree of abatement that may be achieved at a reasonable cost. It appears that regulatory bodies have tried to project the degree of abatement that is "feasible" (that is, not too expensive) and then have promulgated standards incorporating that emission level.[3] This is the rough equivalent of an economic efficiency criterion; the "reasonable cost" reflects an estimate of the benefits to be achieved, and the standard is set to equate the marginal cost and marginal benefit.

One problem in setting this standard or constraint is that the marginal abatement cost curve is not known precisely; estimates may be in error by

3. The federal law actually specifies that the Secretary shall ". . . giving appropriate consideration to technological feasibility and economic cost, prescribe . . . standards. . . ." 42 U.S.C.A. §1857f-1 (a) .

a factor of two for abatement devices now being installed, and more for proposed future devices. If the marginal cost of abatement curve is not horizontal, but has a steeply rising portion (as was found in Chapter 7), then a significant error in setting the emission standard may result in abatement costs far higher or lower than the benefits against which they must be balanced.

Suppose that the marginal abatement cost curve looks like that in Figure 8.1, which is based on Figure 7.2. If the marginal benefit of abatement is about $120 per Pollution Unit, then the optimal degree of abatement is about 80 percent. Yet, if we miscalculate the cost curve and set a standard of 90 percent abatement (an error of only 10 percent), the actual cost may increase to almost $200. This cost curve shape, combined with uncertainty as to its precise location, makes the setting of emission standards quite perilous.

It seems highly unlikely that the government can obtain data on the production costs for abatement devices that are much more accurate than those currently available, unless technological development comes to a halt. This problem of uncertainty will therefore persist and cause continual problems with potential erroneous settings of the abatement level.

The difficulty with the present system is that the information a regulatory body needs to establish proper emission standards is not easily obtained; and, to the extent that it exists, it is in the hands of the automobile manu-

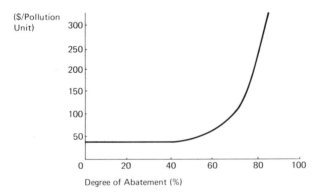

Figure 8.1. Marginal Abatement Cost
Source: Figure 7.2.

facturers and their suppliers. While much useful research on pollution control has been performed by government departments and by various independent firms, the automobile manufacturers themselves are generally in the best position to estimate the production cost of any particular device or engine modification. They have an incentive to be guarded with this information, since disclosure of low-cost abatement devices may lead to stricter standards. If costs are not revealed, the government probably will move slowly and the automobile companies can make modifications at a pace better suited to their production timetables and their regular design changes.[4]

Since standards must be announced a year or two before they take effect, to give all manufacturers time to prepare assembly lines, the required performance may lag a few years behind the best possible performance, as understood by the regulatory agency. One way to avoid this delay is to anticipate the technical progress expected to occur and set standards requiring the incorporation of the best technology at the effective date of the standard. But if the anticipated progress does not occur, some manufacturers may not be able to meet the standard at all; or the costs of compliance may be much higher than those anticipated by the regulatory body. The 1973 relaxation of the 1975 standards is one case where 1970 expectations (or hopes) for technological progress were not fulfilled, requiring an agonizing period of reassessment and finally new standards that could be met for most cars at a reasonable cost.

Aside from the inefficiency of imposing abatement costs much higher than the expected benefits, there are also problems of enforcement. If the standard appears unachievable, manufacturers may make only modest attempts to meet it. They will hardly be worse off with a far miss than with a near miss. A vehicle that is only slightly below the standard can subject the manufacturer to a fine of $10,000 for each automobile, clearly sufficient to stop production. If only one manufacturer met the standard and the penalties were actually imposed on the others, the complying manufacturer would be given a monopoly on automobile production, raising serious antitrust problems. It seems unlikely that the penalties will be imposed

4. For an examination of the behavior of the automobile industry, see Joe S. Bain, "The Technology, Economics and Industrial Strategy of Automotive Air Pollution Control," *Western Economic Journal,* 8, no. 4 (December 1970):329.

unless the vast majority of makes and models meet the standard, since the U.S. economy could not support a prolonged shutdown of the automobile industry. In short, the penalties under the present regulations are so harsh that they have little credibility.

Another problem is that some vehicle parameters, such as weight and transmission type, affect the rate of pollution emissions.[5] The present regulatory system, which imposes the same standard on all vehicles, imposes greater marginal costs on some vehicles than on others. Some models easily meet the standard and can be made still cleaner at low cost. Standards are not the least-cost system for achieving any given reduction in total emissions for the vehicle fleet, since marginal abatement cost is not equal for all vehicles.

PROPOSED SOLUTION The range of uncertainty in abatement costs is thus quite large when standards are used to force technical progress. One mechanism that may cause more efficient and progressive abatement is an effluent charge that levies upon motorists the value of the harm their driving imposes on the community around them. This charge is proportional to mileage actually driven, proportional to the rate of emission per mile, and related to the population density of the area in which the vehicle is used. To minimize the charge, motorists will seek vehicles with lower emission rates and will tend to drive less in heavily polluted areas.

As a practical matter, such a charge is not feasible. Measuring emission rates annually is expensive; it is difficult to record where the mileage has been accumulated and thus what the charge should be. Furthermore, based on total benefits of $8 billion per year, the actual charge will be under 10 percent of average automobile operating costs and will have a small impact (10 percent or less) on total motoring.[6] The maximum justifiable charge, based on $16 billion per year, will reduce motoring by less than 20 percent.

The major impact of the ideal effluent charge is to encourage manufacturers to reduce the emission rates from new cars and to encourage motorists to buy low-pollution, rather than high-pollution, models. In this case, a practical effluent charge can be designed that will have the same impact as

5. See Chapter 5.
6. See Chapter 6.

an ideal charge and will be much easier and less expensive to administer.

The practical solution establishes separately a charge for every gram of carbon monoxide, hydrocarbons, oxides of nitrogen, and lead emitted. When a new make or model of car is introduced, it is tested and its lifetime pollution emission estimated. Then a potential pollution charge is levied upon it, determined by this probable lifetime emission.[7] As production of the model continues, further tests are made from assembly line models and vehicles on the road; any change in emission characteristics for new cars or for the same model after substantial mileage accumulation is reflected by a change in the charge rate. The manufacturer then will have an incentive not just to build clean prototypes but to build durability into his abatement devices and impose quality control on the production line so that the low-emission potential of the design will be achieved in practice.

Measuring emissions by a full California seven-mode cycle is sufficiently expensive that it is hardly worthwhile to measure every car coming off the assembly line and set its tax individually. Instead, the vehicles can be grouped into make, model, engine, and transmission types that have essentially identical emission characteristics. It is necessary to test only a statistically significant sample of each type to estimate the average emissions for that type. A study of the sample will show whether its variance is sufficiently small that it would be fair to levy the same tax on all members of that type. Even if the variance is large, our concern is not with the emissions of individual vehicles, but with reducing average emissions for all vehicles. If the tax is the same on all vehicles of the same design, the manufacturer is motivated to equate marginal abatement costs for each design to the effluent tax. This is the desired efficiency goal.

If total harm from auto pollution were $8 billion per year in 1970, the marginal value of a Pollution Unit of abatement would be $100, assuming linearity. The present value of $100 per year over the life of an automobile is $736 at 6 percent. If a charge had been levied in 1971 based on a rate of $100 per Pollution Unit, the total levy on one vehicle would have been $250, about the same amount as the recently repealed 7 percent federal

7. This may be compared with a scheme proposed by R. d'Arge, T. Clark, and O. Bubik, "Automotive Exhaust Emissions Taxes: Methodology and Some Preliminary Tests," Project Clean Air, Research Project S-12, University of California at Riverside, September 1, 1970.

automobile excise tax. This amount will fall as abatement technology continues to improve. The precise impact of the proposed potential pollution charge will depend on the manufacturers' response to it. Initially, it will create price differentials, lowering the price of cleaner cars relative to dirtier cars. If the manufacturer absorbs differences in charges between clean and dirty cars, he will have an incentive to promote sales of the cleaner cars, because his profit margin on them will then be relatively higher. If the charge is just passed on to the consumer, as seems most likely, then motorists will tend to buy clean cars rather than dirty ones.

Unlike a $10,000 fine, the charge of several hundred dollars is an amount that can be added to the price of a car without disastrous results. In fact, the charge will be levied on all vehicles, in amounts proportional to their emissions. Because the charge is reasonably related to the problem, instead of being punitive and arbitrary, it is entirely credible and even acceptable.

The potential pollution charge thus creates a continuous pressure on the manufacturer to improve his pollution control technology in order to reduce his taxes. No matter how clean his cars, if the manufacturer can make a significant pollution reduction at a moderate cost, he will do so when the potential pollution charge saving is greater than the increase in manufacturing cost. In short, better pollution control is just like a production cost saving for the manufacturer. Comparison of the hypothetical charge with current abatement costs shows that it will lead to meeting at least the 1972 federal standards and cause installation of more advanced devices as soon as cost or performance improves a little beyond current projections. Because automobile emissions have been regulated for less than a decade and the potential for technical progress is great, this incentive effect probably will come close to meeting the 1975 standards without great penalties. Yet it will not force precise compliance if this goal appears enormously expensive.

It may be objected that people should not be allowed to purchase the right to pollute by paying a tax on a dirtier car. This is an argument, not about efficiency, but about equity—an argument that there are some things the rich should not be permitted to do with their money. To satisfy such critics while retaining the advantages of the potential pollution scheme, a maximum emission level may be imposed for which the fee is prohibitive. Since this upper limit will not be the force that causes technical progress, it need only be set at some level known to be achievable with generally available technology. Such a scheme will prevent the sale of cars in which pollution control has clearly been sacrificed for some other objective, such as performance.

While the proposed potential pollution charge is in many ways a radical change from current methods of regulating new car emissions, it can be incorporated easily into the present system. The potential pollution charge can be introduced by an amendment that sets fees for failing to meet the 1975–1976 standards, those fees being the potential pollution charge itself. Thus a sensible charge schedule replaces a preposterous fine. The 1972 or 1973 federal standards can be used as the maximum permissible limit for any vehicle, since most vehicles can already meet them. And if in the future most cars meet the 1976 standard yet growth in automobile use requires further control, the charge can be based on total emissions, not just those exceeding the standard.

The pollution charge can be set with little knowledge of future technological progress or capability. It does not require the impossible; it induces, rather than demands. It imposes on the industry a predictable, constant, and substantial pressure to reduce emissions. This method rewards the industry for every technological advance developed and incorporated, replacing an incentive to argue impossibility with an incentive to do the best possible job. It gives the consumer an incentive to choose cleaner over dirtier cars, something that is entirely lacking now. Such a charge can be incorporated easily into the present legislation, as a penalty for failing to meet the 1975 or 1976 standards, retaining perhaps the 1972 standards as a minimum level of performance. At a later time, it can be modified to encourage improvement beyond the 1975–1976 standards. The federal potential pollution charge should ensure the fastest technical progress in new car emission control without the chaos and uncertainty of the present system. Combined with state and local programs that also can rely in part upon charges in place of standards, it can meet the needs of the most severely polluted areas without imposing excessive costs on the more fortunate.

STATE OR REGIONAL NEW CAR PROGRAM

High-benefit states or regions need some program that will cause production of new cars that emit less pollution than required by the national new car program. This might be a standard of the kind that has been used in the past, but stricter. If, for example, the national program stops with the 1973 U.S. standards, high-benefit states can adopt the 1975 or even 1976 U.S. standards. Alternatively, they can impose an additional effluent charge or potential pollution charge.

The same arguments that favor a potential pollution charge for the federal program apply even more forcefully in the case of high-benefit areas, since they will require more sophisticated technology and thus are the primary beneficiaries of technological change in abatement. A state or regional sales tax on automobiles may be applied proportional to the federal potential pollution charge and in addition to it. This will increase the incentives provided by the federal program and, presumably, result in sales of cleaner cars in such states than in other areas. Even if the federal potential pollution charge scheme is not adopted, states can still impose their own, but they will have to undertake the measurement and calculation necessary to establish such a charge.

While the state potential pollution charge is superior to direct regulation for the same reasons as at the federal level, states have a still better administrative device available: an annual registration fee based on emission rates. The most sophisticated annual pollution charge would be calculated from actual annual emission measurements and annual miles driven, as recorded by a sealed odometer if such a device becomes available. Like the potential pollution charge, this would discourage the purchase of dirty cars, since their lifetime cost would be raised relative to clean cars. It would also discourage *ownership* of dirty cars, which would tend to be sold to buyers in low-pollution areas where the charge was less or zero. It would encourage better maintenance of cars, since a tune-up that saved the owner more on his pollution charge than it cost would provide a positive financial incentive for better maintenance. Finally, it would tend to reduce total mileage driven, since every additional mile increases the annual charge. Because the price is higher for dirty cars, the reduction in mileage would be greatest for the worst polluters. And the low-mileage driver would not be forced to abandon a dirtier car if he drove so little that the charge was not excessive.

The sophisticated annual charge would be appropriate for the worst of the high-benefit areas, certainly including Southern California. The cost of the annual emission measurement is not trivial, however, and the expense of sealing the odometer and ensuring that it has not been tampered with or disconnected may also be substantial. Therefore some high-pollution areas may wish to adopt simpler versions of the charge. One simplification would be to abandon mileage measurement and charge a flat fee based on emissions alone to all cars. This reduces the expense of sealing and monitoring odometers but loses the incentive to reduce motoring and the discrimination between high- and low-mileage drivers.

The second stage of simplification is to drop the emission measurement and base the charge on emissions estimated from sampling cars of each make, model, and year. A relatively small sample of all cars would probably be sufficient to estimate emissions with sufficient accuracy to be equitable. The problem with this system, however, is that the maintenance incentive is lost as soon as the annual measurement is abandoned. The only remaining effect of the tax is to discourage ownership of high-emission makes and models of cars in the state. Still, this may be satisfactory for the less polluted of the high-benefit areas.

If, for some reason, an effluent charge or potential pollution charge is not desired in the state, a regulation can be adopted for new cars. The manufacturers have resisted the adoption of many standards by different states, for the sound reason that they cannot respond to a large number of different conditions at a reasonable cost. Thus the states that desire additional control may have to agree on the appropriate emission level. Given the past history of state cooperation on such matters, particularly when the states are not contiguous, better results probably will be obtained if the federal government sets the higher standard; the states will then have the option of adopting that standard or staying with the one designed for medium-benefit areas.

STATE OR REGIONAL MAINTENANCE, RETROFIT, FUEL
States or regions with low-benefit levels will need no programs of this kind. Those in the upper portion of the medium-benefit range should establish programs that will match the marginal cost of these programs to their marginal benefit level. This will probably include the modification of used cars by installing the manufacturer's kit, which has the same cost per Pollution Unit as new car programs until 1971. It will also include conversion

of fleet vehicles to gaseous fuel operation, which we have shown to impose small costs. Regular maintenance programs and gasoline composition modification are not warranted in medium-benefit areas.

In high-benefit areas, in addition to retrofit and gaseous fuel conversion, alternative methods may be employed, despite their high cost. A regular inspection and maintenance program may be mandatory, despite its high cost, to obtain the small improvement that it can produce. While modifying gasoline composition is much more expensive than installing evaporative controls on the vehicle, this method can be used as an interim measure until the late 1970s; then, evaporative controls will be incorporated into most vehicles on the road.

These programs may be adopted by legislation compulsory for eligible vehicles. If one were certain of their costs and certain of identifying the vehicles for which they are appropriate, then such compulsion would be economically as well as legally sound. It is possible, however, that these things are known only approximately. Who can say, for example, which fleet vehicles will find it not inconvenient to lose half their trunk space and cut their cruising range to 100 miles or less? A blanket requirement will probably miss some vehicles for which it will be no burden and include some for which it creates serious problems. In addition, the cost of gasoline modification is a complex function of the refining process, and may be difficult to determine in the aggregate; it may also vary significantly from one refinery to another.

These problems suggest again that there may be advantages in the use of pricing mechanisms to create incentives where information problems have complicated the job of direct regulation. Maintenance, retrofit, and conversion of fleet vehicles to gaseous fuels can be encouraged to the maximum economical extent by the imposition of an annual effluent charge discussed above as a state new-car policy. Ideally, this would be levied once a year and would be based on actual mileage driven during the year and the actual emission rate for the vehicle, measured at the time of levy. Because of the cost of measuring by a full test cycle as used by the U.S. federal government, the charge would probably have to be based on one of the quick tests, which are less accurate but can be completed in less than one minute. In medium-pollution areas, the charge may be $100 per Pollution Unit emitted, and in high-pollution areas, $200 per Pollution Unit or more. Both will be sufficient to cause motorists to install OLDKIT in cars less than eight years old, and $200 would put it in those ten years old. This will also cause conversion to gaseous fuel by all for whom it would be convenient.

If a tamperproof odometer can be devised, the charge can be based on annual mileage, giving those who drive more and pollute more a greater incentive to clean up their vehicles. Until such an odometer is shown to exist, an annual mileage will have to be assumed; but this can be different for private and commercial vehicles, since the latter have much higher annual mileages.

The use of an annual effluent charge here will have the added effect of tending to encourage the development of technology for reducing emissions of used cars and will encourage buyers to select vehicles with more durable pollution control systems. It will also put pressure on owners of high-pollution vehicles to sell them outside the area in which the charge is levied—that is, in low-pollution areas where their emissions will be less harmful.

The composition of gasoline may be affected by regulation, as has been done in California, or by a charge, levied on the refiner or distributor, based on the composition. The charge will have to be designed to eliminate the undesirable hydrocarbons without placing too great a burden on refineries that cannot adapt or may take several years to do so.

LOCAL MOTORING REDUCTION

The price increases necessary to reduce aggregate motoring are substantial, because of a rather low price elasticity of demand for motoring. Decreasing motoring by 10 percent will require a cost increase of almost $8 billion per year or $400 to $2,400 per Pollution Unit abated, with a loss in consumer surplus of $24–$120 per Pollution Unit.[8] Such a large income effect for a small price effect does not seem worthwhile on a state or regional basis, since other policies have similar costs with much less income effect unless it can be justified on other grounds, such as reducing congestion. If it is possible, however, to cause costs now accrued on an annual basis to be charged on a mileage basis, it will increase the marginal cost without increasing total or average cost. If an odometer that cannot be disconnected or reset is developed, then the cost of insurance and perhaps state registration fees

8. See Chapter 6.

(as well as any effluent charge) can become a function of mileage driven instead of a flat annual fee. If annual costs of $100 can be so shifted (in Boston, insurance alone is likely to run from $200 to $500 per year, so that this may be a minimum), the perceived cost of motoring will be raised by about 10 percent. The decrease in motoring resulting from this change will be about 10 percent, while the cost will be $24 to $120 of lost consumer surplus with no income transfer. Such a policy appears desirable, even on a national basis.

Instead of trying to reduce automobile usage, it is possible to act directly upon ownership on the theory that annual mileage driven per vehicle varies so little[9] that reductions in ownership will be reflected directly in reduced driving. It is clear, however, that most of the country relies primarily upon the automobile for its transportation needs; mass transportation, particularly urban mass transportation, provides adequate service only in a limited number of cities, and there only along a limited number of travel corridors. Any state or regional attempt to reduce automobile ownership by, for example, raising the price will impose an unfair burden upon the many people for whom there is no reasonable alternative. The poor will be the primary sufferers under such a policy; the result will precisely counter recent efforts to increase the mobility of the poor so that they can secure jobs that have been moving increasingly far from urban low-income residential areas. If automobile ownership is to be reduced, it should be done only in local areas where public transportation is known to be an adequate substitute for essential trips such as the journey to work and shopping. Here it may take the form of an annual ownership tax (like that in Massachusetts) or overnight parking restrictions, or both.

Moreover, if the problem is air pollution in the city, reducing automobile ownership in the city may not help much, since much urban traffic consists of cars garaged principally in the suburbs and driven into or through the city during the day. Yet reducing automobile ownership in the suburbs may be difficult, since public transportation within those suburbs is likely to be poor or nonexistent. If reduction in ownership is desired yet no one is to be deprived of essential transportation, an attempt might be made to discourage multiple-car ownership in hopes that suburban commuters to

9. See Chapter 6.

downtown would take public transit to work, leaving the one automobile for trips in the suburbs where transit provides poor service. Those who both live and work in the suburbs would not be able to respond except by car pools. The potential impact of such a program is large, since, of the households owning cars, 37 percent own two or more. A large annual tax on the second and subsequent cars in a household can be the instrument of this policy. There will be difficulties in defining a household for this purpose, but that problem can probably be overcome.

In the long run, we can try to generate new land-use patterns for urban areas, or at least in new urban areas, that will reduce the amount of private transportation needed to perform the usual daily activities. In the short run, it does not appear that we can easily and equitably reduce rates of automobile ownership significantly without serious undesirable side effects except, perhaps, in trying to reduce the rate of multiple-car ownership.

It may be possible to reduce automobile use by improving service or reducing fares on competing modes of public transportation. It appears, however, that this is expensive in terms of dollars per vehicle-mile prevented.[10] If increasing transit ridership costs more than $0.01 or $0.02 per vehicle-mile prevented, then it is a more expensive means of pollution reduction than is further modification of the automobile engine itself. Yet the subsidies to urban commuter railroads and transit systems may run as high as $1.00 or $2.00 per passenger, or several cents per passenger-mile. Only in cases where the subsidy for increased transit ridership is in the order of $0.01or $0.02 per passenger-mile can increased ridership be supported as an economical attack upon automobile pollution. Where the subsidy is greater, other justifications for this expenditure, such as enabling middle-income commuters to live farther from work, must be found.

In areas where existing public transportation service is adequate and where added riders can be carried at a small marginal deficit, it may be desirable to try to restrict driving to that particular destination. Streets can be closed to auto traffic or to through traffic, and street parking can be limited. Higher parking fees may be charged in a particularly polluted area to encourage motorists to use public transportation instead. The costs of policing a parking price increase have already been discussed, and it is probable

10. See Chapter 6.

that total travel to the area will decrease unless the improvement in air quality, noise, and annoyance of cars is so important that it outweighs the added cost of travel to the area. Since, in any city—except perhaps New York—the area well served by public transportation is relatively small, the total reduction in motoring in the metropolitan area also will be small. The real impact of such a program will be on the quality of the air and on traffic congestion in the restricted area itself.

In areas where air pollution levels are high, where public transportation service is good, where a large percentage of travelers already use transit, and where there are adequate bypass roads, automobile traffic may be banned altogether following examples in Sweden. This might be done on a single street, or in an area of several streets, depending on the particular circumstances. Again, unless the improvement in air quality and noise level has a greater impact on people's preferences than the decrease in accessibility, there will be a decrease in total travel to the area. There might be cases, however, where this decrease is thought to be worth the improvement in air quality. It must be remembered that most streets cannot be closed for all time, since delivery vehicles must still have access to commercial establishments, and garbage trucks, police cars, and fire apparatus must be able to get in at any time. This makes it difficult to control such an area. It seems likely that only a moderate number of areas in existing cities will feel that it is worth the many costs of excluding all private automobiles in this fashion.

To summarize, a local motoring reduction program will have several components. On a few streets, or in a small area of the central city that is well served by public transportation, automobile use can be restricted or prohibited, with reductions in pollution only in the affected area. Public transportation service can be upgraded and extended. Parking charges can be examined to ensure that they reflect the full opportunity cost of providing the parking space—that is, to ensure that parking is not subsidized. Taxes may discourage multiple-car ownership in a single household. Insurance fees and perhaps effluent charges can be shifted from an annual to a mileage basis, to increase the perceived marginal cost of motoring. Most important of all, large taxes per vehicle-mile driven could significantly reduce motoring over a complete metropolitan area, although an effective level would require more justification than pollution control. Finally, with a view to the long run, the planning of future land use may be modified to permit a larger percentage of trips to be served well by public transportation or by walking. All these changes except a large regional vehicle-mile tax will have only small or local effects in the short run, but may have a large impact in the long run.

Timing of Emission Controls
It has already been noted that economic analysis alone cannot determine what degree of abatement is desirable, because of the problems of valuing the benefits that accrue. This is a decision that depends on public preferences that are not well defined; and it must be made by the public and its representatives, on the basis of the best available information about costs and effects. It is possible, however, to draw some conclusions concerning the relationship between standards in different places and the timing of such standards.

Let us suppose that the schedule of regulations now established by the federal government is appropriate for Southern California, the area on whose conditions they are largely based. Current estimates are that, if the 1975 standards can be met at all, they will involve marginal costs near $300 per Pollution Unit abated and a total of almost $100 per vehicle per year in abatement costs. While many cities produce photochemical smog on occasion, the frequency and severity of this problem is far worse in Southern California than anywhere else, and the proportion of the population of the country for whom it is a serious problem is no more than 25 percent. If the current federal program is appropriate for Southern California, it cannot possibly be suitable for most of the rest of the country. In medium-benefit areas, the correspondingly correct program would involve significantly lower marginal abatement costs at all times. Again, we recommend a two-level standard if economic waste is to be avoided when costs of abatement rise above the levels of 1971 and 1972.

One response to this problem is to delay the imposition of the expensive standards (those for 1975–1976 at a minimum) or higher pollution charges, in those areas of the country where the automobile pollution problem is demonstrably less than in Southern California. It was shown in Chapter 7 that if the 1975–1976 standards were postponed by ten years, concentrations of HC, CO, and lead would still decrease almost continuously until 1995, despite increases in the vehicle population. Yet if half of the cars in the country could be granted this ten-year reprieve, the savings in capital and operating cost would total $25 billion. Further savings could be achieved by removing the 1973 standards in areas where they are not needed. Since concentrations of all pollutants are now declining, and will continue to do so for almost a decade even if no standards are imposed beyond the 1971 standards, those who feel that in medium-benefit areas the air quality should be cleaner may still be satisfied without further controls. Because of the enormous cost of the 1975–1976 controls, postponing them for at least a few years in medium-benefit areas appears prudent, if not mandatory.

But if it is felt that the 1975–1976 standards are appropriate for the country as a whole, they are hardly satisfactory for Southern California and other high-benefit areas. It can be expected that Southern California will impose the most expensive local and state alternatives found here, including serious restrictions on automobile ownership and use. That Californians have chosen not to restrict motoring, which is admittedly a very high cost alternative, suggests that perhaps the current federal standards may be satisfactory there, after all. In this case, they are clearly inappropriate for much of the rest of the country. If every vehicle built to 1976 standards will cost $500 more over its lifetime than one built to 1973 standards, this standard should not be imposed without conscious consideration that its value equals this cost. This is equally true whether the standard is in the form of a standard or an appropriately high potential pollution charge.

If the federal program should impose controls more stringent than thought necessary in large parts of the country, problems other than massive overexpenditure may result. Most automobile pollution control systems include components that can be removed, disconnected, or adjusted to reduce their effectiveness. These operations also tend to improve gas mileage and power, as well as reducing some severe drivability problems that have arisen in recent vehicles. If motorists in an area are not convinced that the controls are necessary to improve their air quality, the rate of tampering with the devices may rise to a point where the effectiveness of the program is seriously reduced. Effective control may be higher using moderate controls that most motorists believe are desirable than installing more stringent controls that impose such high operating and performance costs that they invite disconnection.

Conclusions

In the early days of pollution control, it was sufficient for the most seriously affected state to establish its own control program and then for the federal government to promulgate uniform standards for the entire country. But as the costs of automobile pollution control rise and promise to jump much higher in the near future, and as pollution control conflicts with energy conservation, more sophisticated strategies are necessary. It is essential that the control program be tailored in some way to the needs of the area if we are to avoid either great economic waste or a revolt against the high costs that might endanger the entire program.

A sensible pollution control strategy must consist of programs at the national, state, and regional level, designed to match the costs of abatement to anticipated benefits for each area. It is both reasonable and necessary that the amounts spent on abatement vary from one area to another in proportion to the severity of the problem. The national program should produce new cars that reduce emissions to a degree needed by the majority of the country, with built-in durability of control systems and at the least possible cost. State and regional programs in medium- and high-benefit areas should encourage or require purchase of even cleaner new cars, where necessary, and may regulate used cars and fuels and even reduce total motoring. These state and regional programs can be designed to deal effectively with the particular problems and conditions in each area.

The best program is one incorporating some form of effluent charge at both the federal and state level. The complete automobile pollution control strategy should include the following:

1. A federal potential pollution charge for all cars, and a minimum emission standard to be met by all;
2. State and regional annual pollution charges in high-benefit areas, based on measured emissions and mileage in the worst of these areas;
3. Fuel tax based on composition;
4. Regional and local motoring reduction program.

The federal and state charges should be phased in over a period of time so that they achieve the desired schedule of abatement without excessive cost or undue delay.

If direct regulation is the primary policy instrument, the complete automobile pollution control strategy will include the following:

1. Federal new car standards—a medium standard applicable everywhere, and a strict standard that may be adopted by states or regions where there is a severe problem, the latter enforced by reasonable, not prohibitive, fines;
2. State and regional retrofit regulations in medium- and high-benefit areas;
3. State and regional maintenance regulations in high-benefit areas;
4. State and regional fuel regulations in high-benefit areas;
5. Regional and local motoring reduction programs.

Either approach can be used to achieve the same degree of abatement, although the speed of technological progress and the cost of the programs strongly favor the effluent charge approach. Some governments cannot act effectively as long as they rely solely on regulatory methods that require them to become experts in complex and rapidly changing technology. They cannot be effective as long as progress requires the regulated industry to tell the regulators how well it can do. And progress will be slow as long as the government must prove what can be done before it can promulgate regulations that will not take effect for two years.

It has been said that the automobile industry has not developed the technology of cleaner cars fast enough to save us from serious pollution problems. If this is true, it is our own fault, for, in fact, the public and the government have failed to produce the innovations in regulatory mechanisms that will induce the best performance by the private sector. A general understanding of what avenues of control are most promising, combined with institutions that will create reasonable incentives to better performance, should produce much more satisfactory results than those achieved with present procedures.

Appendix A Estimation of Fuel Consumption from Gross Vehicle Parameters

Theory

In examining the relationship between some selected gross vehicle parameters and the rate of automobile fuel consumption, we may adopt either
an engineering[1] or an econometric approach. The former begins with
known physical relationships among the components of the system and
builds a detailed model relating the grosser quantities that are of direct
interest. The latter begins with the variables that are of direct interest and
examines large bodies of data concerning them, with the object of estimating their true relationships.

While both approaches are relevant to the problem, the econometric approach will be used for the following reasons: first, it is extremely difficult to
simulate actual driving conditions; second, there are problems in calibrat-
ing a simulation model;[2] and third, it is desirable to reduce the model to
only a few aggregate explanatory variables. The problem remains, however, of securing data that represent typical driving conditions for the average United States motorist, for the rate of fuel consumption per mile depends on the type of trip taken, average speed, and frequency of speed
changes. Moreover, data based on laboratory tests are suspect because the
tests are usually run under ideal conditions of weather, maintenance, and
operation not representative of actual motoring conditions. Therefore, data
gathered from motorists selected at random are preferable, whenever these
are available.

In the American automobile industry, all four manufacturers use a technology of a similar degree of sophistication. Thus it is possible to identify
and measure performance according to a few basic parameters, to which
the details of the technology—transmission design, bearing friction, tire
resistance, and so on—are related in predictable ways. Each of the major
parameters will be considered here in turn.

1. R. K. Louden and Ivan Lukey, "Computer Simulation of Automotive Fuel Economy
and Acceleration," Paper presented to Society of Automotive Engineers Summer Meeting,
Chicago, June 1960, and D. Hwang, "Fundamental Parameters of Vehicle Fuel Economy
and Acceleration," Paper no. 690541, presented to Society of Automotive Engineers Detroit
Section, Detroit, October 30, 1968.
2. Warren has illustrated the dangers of using simulations to predict actual fuel consumption by showing a difference between simulated and actual consumption results. See G.
Warren, "Some Factors Influencing Motorcar Fuel Consumption in Service," Paper no.
65-WA/APC-1, presented at the American Society of Mechanical Engineers, Winter and
Annual Meeting, Chicago, Illinois, November 7–11, 1965.

   Two vehicle parameters that are clearly important determinants of gas
mileage are engine size and vehicle weight. In addition to their direct
relevance, they are highly correlated with other variables. For example, at
high speeds, gas mileage is affected by wind resistance, a function of the
frontal area of the vehicle. This area, however, is highly correlated with
weight. The rear axle ratio is an important factor in fuel economy since it
determines how fast the engine must turn over at a given vehicle speed.
The axle ratio is designed as a function of power and weight; thus in American automobiles—excluding high-performance cars—it has relatively little
independent variation.
   Another variable that has some relevance is the compression ratio. As
the compression ratio is raised in an engine with fixed displacement, the
power obtained per gallon of fuel increases, along with maximum engine
power. In the range of compression ratios currently in use, between 8.5 and
10.5 to one, this relationship can be approximated by a straight line of the
form[3]

$$\text{MPG} = K \times (1 + 0.0625\text{CR}), \qquad\qquad\qquad\qquad\qquad (\text{A.1})$$

where MPG is gas mileage, K is a constant, and CR the compression ratio.
In the regressions discussed later, the compression ratio was frequently not
a sufficiently powerful variable to establish a significant coefficient; so, at
times, it was exogenously included in the form suggested by Equation A.1.
This amounts to a very strong prior belief as to the relevance of the variable
and the form in which it should be included.
   It is not clear, a priori, whether horsepower or displacement should be
the better explanatory variable. In any event, they are highly correlated,
with a correlation coefficient of approximately 0.95, so that they should
give about the same results. The horsepower coefficient should be numerically larger than that for displacement since horsepower is numerically
smaller than displacement.
   Theory does not provide compelling guidance on how the relevant variables should be combined. The effect of power and weight seems likely to

3. See F. W. Kavanaugh, J. R. MacGregor, R. L. Pohl, and M. B. Lawler, "97+ Octane
Fuels Give Best Mileage Economy," *Society of Automotive Engineers Journal,* 66 (October
1958):28.

be multiplicative rather than additive, since in traffic more weight provides an opportunity to utilize more power without reaching excessive speeds. The compression ratio, when used, reflects an ability to obtain more power per gallon of fuel and therefore should be multiplicative. A linear form, however, would be easy to use for quick calculations, and might be justified if it did not have a low R^2 or highly autocorrelated residuals. The linear form is

$$\text{MPG} = C + a \times \text{WT} + b \times \text{HP (or DIS)}, \qquad (A.2)$$

where WT is weight in pounds, HP is horsepower, DIS is displacement in cubic inches, and C is a constant. The multiplicative form is

$$\text{MPG} = \frac{C}{\text{WT}^a \times \text{HP}^b} \qquad \text{(or DIS).} \qquad (A.3)$$

Adjusting for compression ratio, we solve Equation A.1 for K and substitute this for MPG:

$$\frac{\text{MPG}}{1 + 0.0625\text{CR}} = \frac{C}{\text{WT}^a \times \text{HP}^b} \qquad \text{(or DIS).} \qquad (A.4)$$

This is estimated logarithmically as

$$\log \frac{\text{MPG}}{1 + 0.0625\text{CR}} = C + a \log \text{WT} + b \log \text{HP}. \qquad (A.5)$$

It is useful to compare these equations and variables with those used by Fisher, Griliches, and Kaysen[4] in an earlier study of the problem. The authors sought to identify a relationship between gas mileage and a fuel economy factor, F, which was defined as a constant times the number of engine revolutions per mile times the displacement. Such a formulation includes weight only indirectly in F; since, for a given engine displacement,

4. F. M. Fisher, Z. Griliches, and C. Kaysen, "The Costs of Automobile Model Changes since 1949," *Journal of Political Economy*, 70, no. 5 (October 1962): 433–451.

higher axle ratios will be used in heavier vehicles to preserve adequate acceleration, and the number of revolutions per mile will rise, raising F. This Fisher study therefore did not separate the effects of power and weight, nor did it include weight as a separate variable. If weight is an important variable, one would expect higher R^2 where it is explicitly included, which is the case in the present study.

Data

To study the relationship between gross vehicle parameters and fuel consumption, we will use three sets of data. Two are from gas economy runs sponsored by major oil companies and conducted under carefully controlled conditions. They are not necessarily representative of actual driving conditions but provide relatively precise measurements for the driving cycle that they represent. The third is from surveys of owners of automobiles, selected at random from all owners of the particular makes and models surveyed. In this case, the driving cycle is probably an excellent representation of the national average, but there is some question about the precision of the mileage figures supplied by motorists in response to such a survey.

PURE

The Pure Oil Performance Trials are run annually at Daytona, Florida, and represent a controlled scientific experiment. For this study, the trials of 1966 and 1967 are used, consisting of 47 and 59 vehicles, respectively, ranging from luxury cars to compacts and sport compacts. The test results specify make and model, and engine displacement and horsepower.[5] A number of identical pairs of vehicles are included each year, so that it is possible to see how much variation in the gas mileage can be attributed to random manufacturing differences or to random differences in the tests themselves. The variation averages less than 0.5 MPG or 3 percent. Given the driving cycle used, these data seem to be excellent in precision of result, knowledge of all variables, and similarity of conditions for all cars.

MOBILGAS

The Mobilgas Economy Run was an annual test run across the United

5. Complete test results are cited in *Automotive Industries*, March 15, 1966, 1967. Only vehicles with automatic transmissions were included, since 90 percent of all cars built in the United States have this option. The Corvair was excluded from this and other data sets since its air-cooled, horizontally opposed engine employs a somewhat different technology than all other domestic engines.

States every spring until 1968. The cars were stock models, as in the Pure Oil Trials, and were carefully tuned. The Run was on public streets and roads, but the average speed was in the vicinity of 50 miles per hour; thus it was clearly representative of long-trip driving rather than urban use. All cars had automatic transmissions, and between 40 and 50 ran every year. For this study, the runs of 1966, 1967, and 1968 are used, including a total of 131 observations.

Since no specifications were given for the entrants, other sources are used for engine size and vehicle weight.[6] This procedure introduces a potential source of error in the independent variables, since most domestic automobiles are offered with a variety of engine sizes, even in a single make and model.

SURVEY

These data come from a series of owner surveys conducted by *Popular Mechanics* magazine in which a random sample of owners of new cars is asked, "What gas mileage are you getting? Local driving —— mpg. Long trips —— mpg." The owners also state the number of miles they have driven their car at the time of the survey. The survey results presented include the median gas mileage reported and, occasionally, a histogram showing the range of mileages experienced. An average of 300 owners responded to each survey; thus even if individual estimates of mileage are not precise, the median should be, unless there is some consistent bias in the results.

Because the total mileage reported by owners varies from one survey to another, it appears that the precision of the gas mileage data is subject to some variation among vehicles surveyed. Thus the data are weighted by the square root of the number of miles reported, which is proportional to the number of returns received.

Results

Since there are differences among the three data sets that can be expected to affect the results of the regression analysis, separate regressions have been run on each set. The results are shown in Table A.1. The first four results are from simple linear equations, in the form of Equation A.1. The

6. *Automotive Industries,* vols. 134, 136, 138 (March 15, 1966, 1967, 1968).

Table A.1. Gas Mileage Regressions

Data Source	Equation No.	Dependent Variable	Unadjusted Coefficients				Coefficients Adjusted to 14-15 MPG				
			C	CR	WT	DIS or HP	R^2	C	CR	WT	DIS or HP
Pure	Linear a	MPG	24.3	1.136 0.262	−0.00260 0.00043	−0.0237 0.0031 DIS	0.87	19.45	0.908	−0.00208	−0.019 DIS
Mobilgas	Linear b	MPG	28.7	—	−0.00082 0.00032	−0.0208 0.0024 DIS	0.80	23.0	—	−0.00066	−0.0166 DIS
Survey	Linear c	MPGT	28.3	—	−0.00178 0.00060	−0.0217 0.0037 HP	0.84	28.3	—	−0.00178	−0.0217 HP
	Linear d	MPGL	25.4	—	−0.0016 0.00055	−0.0243 0.0033 HP	0.88	25.4	—	−0.00160	−0.0243 HP
Pure	Log e	LAMPG	8.49	—	−0.532 0.088	−0.299 0.045 HP	0.89	8.26	—	−0.532	−0.299 HP
Mobilgas	Log f	LAMPG	6.34	—	−0.299 0.053	−0.344 0.082 DIS	0.86	—	—	−0.229	−0.344 DIS
Survey	Log g	LAMPGT	6.59	—	−0.273 0.109	−0.370 0.043 HP	0.91	—	—	−0.273	−0.370 HP
	Log h	LAMPGL	7.10	—	−0.290 0.116	−0.473 0.046 HP	0.93	—	—	−0.290	−0.473 HP

Note: The lower number is the standard error for the coefficient. MPG is miles per gallon, all driving; MPGT is trip mileage; MPGL is local mileage; LAMPG is the log of miles per gallon, all driving; LAMPGT is the log of trip miles per gallon; LAMPGL is the log of local miles per gallon; C is the intercept; CR is the compression ratio; WT is gross vehicle weight in pounds; DIS is displacement in cubic inches; HP is power in horsepower.

WT coefficient is the change in gas mileage from a 1-pound weight increase, and the DIS or HP coefficient shows the change in miles per gallon for an increase in engine size of 1 cubic inch or 1 horsepower, respectively. The second four are logarithmic forms for the relationship of Equation A.2, so that the coefficients show the percentage of change in gas mileage for a 1 percent change in weight and engine size. In Table A.1, Equations e through h include the compression ratio adjustment explicitly; b through d ignore it, and a estimates the coefficient for the compression ratio separately. The first is the only linear form in which the compression ratio is strongly significant. Theory favors the logarithmic forms, but the fit of the linear forms is reasonable, and they are much easier to use in simple computations. Table A.1 contains both raw regression results and results adjusted to a common average fuel consumption.[7]

One way to view these results is to assume that the survey results are the most realistic, since they are based upon actual driving conditions, and then to ask whether the mileage test results, given the special conditions under which they were run, contain any factors that invalidate the survey results. The Mobilgas results seem to confirm the survey results, since the engine size coefficients are similar and the smaller weight coefficient in Mobilgas should result from the steady speed conditions of that test. Only the high

7. The average gas mileage reported in the two economy runs is well above that observed in the user survey. The survey vehicles have average characteristics close to the national automobile fleet average, and they record an average local gas mileage of 13.9 MPG and a trip performance of 16.8 MPG. The Department of Transportation in *Highway Statistics 1968* (Washington, D.C.: U.S. Government Printing Office, 1969), Table 1, reports average gas mileage for the U.S. at 13.79 for 1968. It is estimated that the proportion of local and trip driving, as these terms are used in the survey, is about two to one, so that the average gas mileage in the survey is just under 15 MPG. This is only 7 or 8 percent better than the DOT figure, probably because survey cars are only about one year old, and therefore is not adjusted. The average mileage for the Pure Oil Trials is 18.56 MPG. Noting that the average parameter values for the automobiles in the Pure Oil Trials are different from the average for all U.S. automobiles, we can use the estimated power and weight coefficients to compute what the average Pure Oil gas mileage would have been if those cars had the average U.S. characteristics. The estimated mileage is still 20 percent above the U.S. average. Assuming that the factors which make Pure Oil gas mileage high operate proportionally on all automobiles and are independent of the vehicle parameters such as maintenance and adjustment, skillful driving, optimum operating conditions, and a different driving cycle, we have recomputed the coefficients for Pure, after reducing all gas mileage figures by 20 percent. The resulting coefficients are listed in Table A.1 as Coefficients Adjusted to 14–15 MPG. Since the average mileage in the Mobilgas Economy Run was 19.56 MPG, similar adjustments have been made to produce an average of 15 MPG.

weight coefficient in Pure remains unexplained, since one would expect a mileage run to be less sensitive to weight than is actual driving. Furthermore, the low standard error for the weight term in Pure and the high standard error in the survey suggests that the Pure finding cannot be easily dismissed. Further examination of the data for the two runs shows that the survey has a range of weights slightly smaller than Pure, and the standard deviation of weights from the mean is 20 percent smaller for the survey. The smaller variation in the independent variable may account, in part, for the lesser confidence level of the estimate on that variable.

In general, less credence is given to the Mobilgas equations than the others. This interpretation is justified, first, because of the potential errors in the engine size variable, referred to in the data section and strongly hinted at by the R^2, which is always the lowest of the set. This probable error in an independent variable could cause bias in the parameter estimates and—more important for a large sample such as this—inconsistent estimates. Second, the Mobilgas run is the closest of all to long-trip driving under ideal conditions and does not include significant amounts of the urban driving that comprises the large part of the typical motorist's use of his car.

If the Mobilgas results are set aside, the most appealing single equation is that from the survey, for trip conditions, Equations c and g in Table A.1. Here the engine size coefficient is between those for Pure and those for local driving; this calculation follows from Cornell's finding that high-speed fuel consumption is less sensitive to power than local driving.[8] In the linear form, d, the weight coefficient −0.00178 is between that of Pure −0.00208 and local driving −0.0016; while, in logarithmic form, the local and trip coefficient for weight −0.290 and −0.273 are so close as to be indistinguishable. What is puzzling is that the weight coefficient for Pure is above those for the survey, and by a large amount in the logarithmic form. One would expect that a mileage run would be less sensitive to weight than actual conditions, if only because total variation in gas mileage should be less with careful driving than with everyday driving.

Linear Survey Equation c implies that a weight increase of 1 pound will

8. J. Cornell, "Passenger Car Fuel Economy Characteristics on Modern Superhighways," Paper no. 650862, SAE National Fuels and Lubricants Meeting, Tulsa, Oklahoma, November 2–4, 1965.

reduce gas mileage by 0.00178 miles per gallon, while a power increase of 1 horsepower will decrease gas mileage by 0.0217 miles per gallon. With average weight and power of 3,700 and 250, respectively, a 10 percent reduction in gas mileage is effected either by raising weight 800 pounds (21 percent) or raising power 65 horsepower (25 percent), giving elasticities of 0.48 and 0.40, respectively. The logarithmic form of this equation suggests elasticities of 0.27 and 0.37 for weight and power.

Table A.1 shows the similarity between the coefficients estimated from the mileage tests and the owner survey. The weight coefficients in the survey are always between those of the two mileage tests, and the engine size coefficients are not far above them. This suggests that, despite the great difference in source of data, there is a strong underlying relationship that is captured in all the results. If the displacement coefficients are increased by about one-third to make them comparable to the horsepower coefficients, then the survey results all fall between the mileage test results.

Another interesting finding is that the engine size coefficient usually is substantially more certain than the weight coefficient. In all but the Pure Oil equation, the weight coefficient is significant only at slightly better than the 2 percent confidence level, while the engine size coefficient is always well beyond the 1 percent level. This represents in part the greater variation in engine size, by a factor of almost 3, than of weight, a factor of 1.6. That the excess in explanatory ability of engine size is greatest in the survey data and smallest in the Pure Oil Trial shows that a large engine does not necessarily mean worse gas mileage in the hands of a careful professional driver but that, in the hands of an average motorist unconcerned with fuel economy, it raises a greater potential for increased fuel use.

Scheffler[9] used a linear model for a city driving cycle and found an increase of one MPG from a 400-pound reduction, one MPG from a 120-horsepower reduction, and one MPG from a 60-degree rise in temperature. This result corresponds to coefficients of -0.0025 for WT and -0.0083 for HP, which is larger than the largest weight coefficient found in the present study, and less than one-half of the smallest power coefficient. Scheffler states that city driving is more sensitive to power than is highway driving;

9. C. Scheffler and G. Niepoth, "Customer Fuel Economy Estimated from Engineering Tests," Paper no. 650861, SAE National Fuel and Lubricants Meeting, Tulsa, Oklahoma, November 2–4, 1965.

this finding is consistent with the present survey results (the power coefficient is larger for local than trip driving) but makes his small power coefficient for city driving even more surprising. The insensitivity to power could result, in part, from the use of warmed-up vehicles, since a cold engine uses considerably more fuel than a warm one and a large engine warms up more slowly than a smaller one.[10] It may also reflect some important difference between the maintenance of the proving ground vehicles and the survey's vehicles, between the proving ground driving cycle and actual owner driving patterns, or between the composition of the proving ground and survey fleets. Without further information, it is not possible to reconcile this attribution of small significance to horsepower with the present results.

10. Scheffler concludes: "City fuel economy engineering test data would more closely parallel customer experience if based on a driving schedule that included warm-up." Ibid., p. 4.

Appendix B Valuation and Demand
for Vehicle Attributes

Determination of Attribute Prices

Many goods, particularly consumer durables, are offered in a variety of models, with different prices for each model. The physical or technical characteristics or attributes that distinguish these models may influence the cost of manufacture or the price consumers are willing to pay. These attributes may be continuous physical parameters, such as weight, length, or volume, that can be easily observed. They may be discrete physical parameters, such as the number of doors, presence or absence of certain accessories or features, or voltage capacity. They may also be parameters that cannot be observed directly, such as reputation for durability, economy of operation, and availability of reliable service. For any good, we can propose a list of attributes thought to be the primary bases upon which consumers choose among available makes and models, and which therefore influence the price of each. In some cases, it may be desirable to use an attribute that is not particularly valuable in itself but is easy to measure and highly correlated with attributes of some basic appeal. Weight, for example, may be of little intrinsic value in a refrigerator but may correlate well with interior volume and durability.

The hypothesis presented and tested here is that the market price of domestic new cars is related in a predictable way to physical dimensions of the vehicle, such as weight, power, and length. Thus, the price of a model is given by

$$P_j = f(X_{1j}, X_{2j}, \ldots, X_{nj}), \tag{B.1}$$

where each X_{ij} is the quantity of attribute i possessed by model j.

Griliches used this hypothesis to find the value placed by consumers on changes over time of these physical dimensions of vehicles.[1] Our purpose in testing the hypothesis is to provide a basis for estimating welfare losses resulting from constraints that might be imposed on certain of the dimensions. This requires some further investigation, on both the demand and

---

1. Zvi Griliches, "Hedonic Price Indexes for Automobiles: An Econometric Analysis of Quality Changes," in *The Price Statistics of the Federal Government*, National Bureau of Economic Research, 1961. The theory of hedonic pricing is explained in this work, and will not be repeated here.

the supply sides of the market, of the significance of any relationship found between parameters and prices.

Two forms of the relationship are tested here, a linear form

$$P = a + b_1X_1 + b_2X_2 + \ldots + b_nX_n,$$ (B.2)

and a semilog form

$$\log P = a + b_1X_1 + b_2X_2 + \ldots + b_nX_n.$$ (B.3)

Both perform well. A good fit for Equation B.2 on a set of price and attribute data means that prices rise linearly with each of the attributes tested, and that the price function is separable in that the marginal value of an attribute is independent of the levels of the other attributes. Furthermore, with this linear formulation, the marginal and average (per unit) values of an attribute are the same and constant over the relevant range.

Our model is based on the usual economic assumption that utilities of different individuals are independent of each other. It may be argued that, for a visible consumption good like an automobile, this assumption is particularly vulnerable. People may wish to have the most powerful car on the street, regardless of absolute horsepower values; this would suggest a demand function that is highly relative to what others are buying. If this is the case, then the cost of policies that lower power or raise its cost will be less than that found here. Space does not permit a detailed analysis of the relative power hypothesis; but our analysis will assume independent utilities, and the results may be adjusted downward to the extent that the true value is felt to be relative and not absolute.

The number of a single-attribute good sold at every price represents the number of people for whom the attribute in that quantity is worth at least the price, but not as much as the next-higher price. The loss of consumer surplus incident upon a change in the supply curve is the difference between what consumers pay afterward for various attribute levels and the amount earlier discovered to be their value. The demand curve for each attribute can be found by observing sales when the price of the attribute is varied and all other attributes and prices are fixed. Alternatively, we can use multivariate regression analysis of price on the attribute quantities of a variety of different attribute bundles. This works, however, only when the values of the attributes are mutually independent. If the consumers' utility functions are not separable in the attributes, or if these utility functions are not separable among consumers, then there is no simple solution to the problem of defining and identifying the price curve for each attribute. Here we assume that utility functions are separable in consumers and attributes. This is probably a fair approximation, and it is analogous to the usual assumption of independent preferences of consumers.

The relative importance of different models can be reflected by weighting observations by the number of units of each model actually sold. The prices of popular models will thus exert more influence over the results than those of less popular models.

In a perfectly competitive industry, the price for each model will just equal the zero profit price, and all prices will precisely display manufacturing marginal costs. It appears that this is not a valid assumption for the U.S. automobile industry, which is characterized by giant firms that follow the price leadership of one—General Motors. It does seem safe, however, to assume that manufacturing cost should rise monotonically with various size parameters, and the presence of four competing firms should prevent gross disparities between cost and price. If the price is not a perfectly competitive price, earning zero profit on each model, any nonmarginal change in the mix of models sold may cause changes in producers' surplus. The cost data are not sufficient for us to evaluate this change, however, and we will have to assume that relative prices reflect relative costs, as though the prices used were set under conditions of perfect competition.

The parameters measured here are chosen to represent attributes in which purchasers have a direct or indirect interest and which are easy to measure. Intuitively, it seems that automobile purchasers are interested in interior space not only for passengers but also for cargo, peformance, economy, comfort, prestige, and the convenience offered by various accessories. Here length has been used as a general proxy for size and space. Somewhat more information probably could be added by using a roominess index, but the high degree of multicollinearity between such an index and length would minimize the amount of information added and would make accurate estimation of both coefficients difficult or impossible.

If other things such as handling characteristics and type of suspension are held constant, the quality of the ride (passenger comfort) increases with increasing vehicle weight and a longer wheelbase. Since wheelbase is highly correlated with overall length, it is already accounted for. In addition to this demand for weight in itself, there is a derived demand for weight arising from the fact that many accessories or other qualities, such as durability, are necessarily accompanied by an increase in weight. Therefore value should increase with weight, for both direct and indirect reasons.

An automobile's performance, or rate of acceleration, depends upon its weight and horsepower. Another factor related to engine size is displacement, which specifies the physical volume of the engine as opposed to the amount of power it produces. Displacement is of interest because, for a given power level, larger displacement means that the engine is not working as hard and may last longer. Also, obtaining high power from a larger displacement engine generally means that the engine is less sophisticated and thus easier to maintain. Finally, for a given power level, a larger displacement usually means a lower compression ratio, which permits use of less expensive fuel. Thus, while horsepower and displacement are highly collinear, they are not interchangeable, and both have been used separately in this study.

For a given type of driving, the operating cost of an automobile depends upon its power or displacement, weight, and durability. The first three of these attributes are already included; presumably, their evaluation will include some allowance for fuel economy. This raises the question of whether people shopping for an automobile consider the future stream of operating costs and, if so, how. For our purposes, it is not necessary to decide whether the consumer considers the present value of the discounted future cost stream or applies a less sophisticated analysis as long as he accounts for the operating cost in some systematic way. The demand for power and weight will be assumed to include adjustments for whatever allowance the consumer gives to these future operating costs.

Another characteristic that probably influences manufacturing costs and consumer preferences is whether the engine has six or eight cylinders. In the minds of consumers, there may be some difference between the two engine types, other than the usual power differences; this may lead to willingness to pay different amounts per horsepower for them. Thus a dummy variable has been used to indicate whether or not the engine has eight cylinders.

Styling is an important factor in determining market share in the automobile industry, if not market price. There is, however, no single measure of styling that can be applied to the various automobiles produced. Therefore two body types have been studied here, combining a style feature with a convenience feature: the two-door hardtop (the most popular model) and the four-door sedan. The first has been used for most of our analysis, and the second in cases where a particular vehicle is not available in the two-door hardtop or when more models have been needed to achieve proper weighting for the make. Generally four doors cost more than two, and a hardtop costs more than a sedan, so that the relative prices of a two-door hardtop and a four-door sedan cannot be determined a priori. A dummy variable has been included for body type to determine whether there is a significant difference.

Finally, it seems likely that there are luxury or accessory differences among automobiles that are not covered by the above variables. Interiors can be finished in materials of different quality without affecting weight or any other parameter, yet affecting the selling price of the vehicle. Many convenience items can be added—such as a clock, special interior lighting, power seats, power windows—that have little weight yet indicate the degree of luxury of the automobile. While the basic vehicle costs about one dollar per pound, a radio weighing only a few pounds may cost $50 to $100, or more than $10 per pound. Thus a quality variable has been added that relates to these identifiable accessories that are standard equipment and to the status the model occupies in the maker's product line. All accessories are included except automatic transmission, power steering and brakes, and air conditioning. On the few models where automatic transmission, power steering and brakes, and air conditioning are standard equipment, their price has been subtracted from the list price, and the automobile treated as if they were not present.[2] For the dozen or so models where they are standard, the amount subtracted from the price is the highest price charged by

2. Griliches and Triplett have found the presence of an automatic transmission generally insignificant in its effect on price, although as an option it costs $150 to $225, and on used cars it raises the value by $100 or so. Power steering and brakes, on the other hand, raise price by over 20 percent, clearly reflecting much more than the production value of this $50 option. Griliches, "Hedonic Price Indexes for Automobiles"; and Jack E. Triplett, "Automobiles and Hedonic Quality Measurement," *Journal of Political Economy*, 77, no. 5 (1969):408–417.

that manufacturer for a given option, since it costs more on expensive than on economy cars.

DATA

The year 1968 was selected for this study because it is sufficiently recent that technology and product mix have changed little up to the present, yet sales figures are now complete and there is some history of depreciation rates from used-car prices. For a given make and model, determining the number of cylinders, body style, and length has caused no problem, but estimation of weight has been more difficult.

The data consist of 189 observations on new car list prices for automobiles manufactured in the United States.[3] It is true that automobiles are usually discounted from the list price, but there is not sufficient information on the extent and amount of this discounting to permit its incorporation into the analysis. It is assumed that the extent of discounting must not vary greatly among the four manufacturers, and that if it varies with price, it is in some predictable and systematic fashion, such as in proportion to price. Thus discounting simply means that the coefficients estimated here are too high by a common amount.

As far as possible, the number of models chosen to represent each make are in proportion to the number of units of that make sold in 1968. This is done to incorporate volume into the analysis by ensuring that where a price is too high, causing low sales, that price will have little influence. Two exceptions to this weighting are American Motors, which is represented by a more than proportional number of models in order to provide a representative sampling of the automobiles it offers, and Ford, which is underrepresented, as there are not enough models to reflect its high sales.

All models used are standard models with no optional equipment, except that every automobile which has a standard six-cylinder engine also appears with the smallest standard V-8 engine. This has several implications for the results. First, only the smaller engines are represented here. The large optional engines do not appear except in the few cases where a sep-

---

3. Data on length are taken from *Automotive Industries,* March 15, 1968, p. 153; on weight and power, from the same at pp. 133–134; and on quality, from the table of convenience items on p. 154. Price is from the *Automotive News 1968 Almanac Issue,* pp. 55 and 57. Price is as of April 1, 1968, and includes the 7 percent federal excise tax and suggested dealer preparation charge. Transportation, state and local taxes, and options are extra. The heater alone is standard.

arate model is offered with the large engine, as in the Chevrolet Chevelle SS396 or the Plymouth Roadrunner (intermediate-size cars that come only with a very large engine), or in the case of luxury cars that have enormous standard engines. Thus, if the marginal cost of added power is not a linear function of power, this study captures the cost at lower price levels, but not at high power levels.

A separate examination has been conducted on the pricing of optional medium-power engines for low-price cars. The data used are the extra cost of optional engines over the cost of the standard V-8 engine.[4] Since no information is available on the weight of these engines, we have assumed that, once a V-8 engine is used, a larger one will not change the weight substantially. This is certainly not true for the largest engines, which usually require the purchase of heavy duty suspensions and drive trains, but it may be accurate for the middle range of engines.

Because most luxury options are not included here, except in the luxury cars on which they are standard, this study does not indicate the impact these items have on the price and weight of automobiles for which they are purchased. In general, such items are not very heavy, however, so they should have little effect on the price-weight-power relationship that is our present focus.

RESULTS FOR 1968

Several equations have been fitted using least-squares regression techniques, and some provided both high R^2's and quite significant coefficients for all variables. Four equations have been selected as representing the best possible fit on the data; two of them are linear and two semilogarithmic. The summary statistics for these four, a fifth version that uses somewhat different variables, and results from Griliches and Triplett are shown in Table B.1. Three of the equations use horsepower and two use displacement as the engine size parameter. Since these two variables are highly collinear, their fit is nearly identical; both are presented because application of these results may call specifically for one or the other.

Of the first two equations, it appears that the second, using horsepower rather than displacement, is slightly preferable, with its higher R^2 and somewhat lower standard errors. Similarly, Equation 5 is slightly better

4. From the National Automobile Dealers Association, *Official Used Car Guide,* April 1968, Central Edition.

Table B.1. Regression Results on 1968 Prices[a]

Dependent Variable	Const	WT (lb)	LNG (in.)	DIS or HP	QUAL	DCL	DBY	R^2
1 Price	521 119	0.469* 0.050	—	2.32* DIS 0.35	54.7* 8.5	−126* 37	−76.5* 22.9	0.886
2 Price	615 117	0.509* 0.043	—	2.20* HP 0.28	47.9* 8.2	−104 32	−76.1* 22.1	0.894
3 Price	523 339	0.484* 0.106	1.05 3.000	1.69* HP 0.29	49.6* 8.4	—	−76.3* 22.8	0.887
4 Log Price	7.16 0.04	0.000167* 0.000016	—	0.000665* DIS 0.000110	0.0179* 0.0027	−0.0301 0.0119	−0.0304* 0.0073	0.896
5 Log Price	7.19 0.04	0.000179* 0.000014	—	0.000622* DIS 0.000091	0.0160* 0.0026	−0.0233 0.0104	−0.0303* 0.0071	0.901
6 Log Price[b]		0.000136* 0.000046	0.0015 0.0017	0.00119 DIS 0.00029		−0.039 0.025		0.951
7 Log Price[c]		0.000221 0.000014						0.891

WT = curb weight in pounds; LNG = overall length in inches; DIS = engine displacement in cubic inches; HP = engine maximum-rated horsepower; DCL = 0 if 6 cylinders, 1 if 8; DBY = 0 if 2-door hardtop, 1 if 4-door sedan; QUAL = standard accessories.

*Significant at 1 percent.

a. Standard error is below coefficient.

b. Zvi Griliches, "Hedonic Price Indexes for Automobiles: An Econometric Analysis of Quality Changes," in *The Price Statistics of the Federal Government*, National Bureau of Economic Research, 1961, Table 3, 1960 data.

c. Jack E. Triplett, "Automobiles and Hedonic Quality Measurement," *Journal of Political Economy*, 77, no. 5:408–417, Table 2, 1967–1965 data, truncated model.

than Equation 4, again because of the use of horsepower rather than displacement. The first three equations use different forms of the dependent variable, price, than the last two, so that their explanatory power and R^2's are not comparable.

A surprising result is that in most linear equations in which length is included, its coefficient is both insignificant and negative, while log-linear forms give an insignificant but positive value. Table B.2 suggests that vehicle length is well explained by weight, and it is not as clearly related to price.

Weight always has the greatest degree of confidence, with a t-statistic over 4.0 when length is in the equation, and near 10.0 when length is omitted, significant well beyond its percentage level. Equation 2 shows that extra weight is valued at approximately $0.509 per pound, and this value changes little, even in Equation 3 when the highly collinear length variable is added. Griliches[5] has found that 100 pounds raises the price by 1.4 percent. For a $3,000 automobile, this would be $42, or $0.42 per pound —very close to the 1968 result obtained here. The log coefficient corresponding to Griliches's 0.136 for weight is 0.179 from Equation 5. The higher values found by Triplett[6] are probably related to the failure of his power coefficient to explain consistently a significant portion of the price.

Griliches's log coefficient for power in Equation 6 is nearly twice that of Equation 5, giving him a value of $3.60 per horsepower in 1960. This drop in value, over eight years, may reflect a real decrease in the marginal cost

Table B.2. Correlation Coefficients

	LNG	WT	HP	DIS	QUAL	PR
LNG	1.0	0.91	0.53	0.57	0.59	0.77
WT		1.0	0.76	0.80	0.65	0.89
HP			1.0	0.97	0.53	0.82
DIS				1.0	0.51	0.82
QUAL					1.0	0.73
PR						1.0

LNG = overall length in inches; WT = curb weight in pounds; HP = rated engine horsepower; DIS = engine displacement in cubic inches; QUAL = a measure of accessories and luxury; PR = price.

5. Griliches, "Hedonic Price Indexes for Automobiles."
6. Triplett, "Hedonic Quality Measurement."

of power; it seems likely, however, that it comes in part from differences in the sample data. First, the current data are weighted for sales, and Griliches's were not. In addition, the quality variable used here may explain some price variation that is collinear with power, and therefore contained in Griliches's coefficient.

This result must be interpreted in the light of the cylinder coefficient (DCL), whereby, at a given power level, the substitution of an eight-cylinder engine for a six will *reduce* the price by $104. Once again, this is counterintuitive, since people generally pay more for an eight than for a six. There is, however, no overlap between the power levels offered in the two kinds of engine; thus, if a motorist wanted the power of an eight in only six cylinders, he would have to pay a premium for it. This conclusion was reached by Griliches also.[7] Since the power levels of the two engines do not overlap, the production cost curves may be nonlinear, and it may not be more economical to build six-cylinder engines for small power levels.

Triplett[8] disputes the validity of a power coefficient, stating that his power and cylinder coefficients are insignificant and that the weight coefficient alone can account for the difference in price between the two engine types. Two reasons for this result may be suggested. First, purchasers may care only about weight and not power, so that extra horsepower has no value except as it adds weight to the vehicle. This is clearly not true, as manufacturers advertise power very strongly and rarely mention weight.

The other explanation may be that purchasers care about power but do not care about weight except as it relates to other features such as accessories, durability, power, and space. If this were the case *and* if the relationship between power and engine weight were such that one dollar's worth of power weighed just as much as one dollar's worth of accessories, durability, and space, then the weight variable alone could explain the price of different engines. It would be entirely fortuitous if this relationship should hold, however; it is more likely that the value of power would not be precisely collinear with the value of weight from other causes. In this case, separate coefficients could be found for weight and power; the value of a larger engine would be the power coefficient *plus* the weight coefficient (assuming that the added weight gave nothing desirable except power),

7. Griliches, "Hedonic Price Indexes for Automobiles."
8. Triplett, "Hedonic Quality Measurement," p. 410, n. 4.

or the power coefficient alone, if the extra weight were thought to make the engine more durable. In either event, a significant power coefficient would be expected.[9]

The coefficients here are for standard six-cylinder or eight-cylinder engines only. We have conducted a separate study of intermediate size optional V-8 engines to see whether the estimated results continue to hold. The results are shown in Table B.3, where an average marginal price per horsepower seems to vary around $1.40. The asterisks indicate engines so large that they usually require special transmissions, suspensions, and drive trains, all of which can add considerable weight. All other engines are only moderately larger than the standard eight and add only a small amount of weight. Thus the marginal cost of $1.50 is actually a coefficient for both power and weight, and the value of power alone should be somewhat lower. This marginal cost per horsepower is below the estimated coefficient, and the marginal cost per cubic inch is also below the estimated $2.32. This

Table B.3. Marginal Cost of Optional Engine Size

Make	Model	V-8 Base Engine		Optional Engines		$ Over Base	$/HP	$/cu inch
		DIS	HP	DIS	HP			
AM Motors	All	290	200	290	225	96	3.24	0.00
				348	280	117	1.47	2.01
Ford	Fairlane	289	195	302	210	26	1.73	2.00
				390	265	97	1.38	1.01
				390	325	177	1.36	1.75
	Mustang	289	195	302	230	66	1.88	5.07
				390	325	158	1.21	1.56
	Ford	302	210	390	265	78	1.41	0.88
				390	325	158	1.37	1.79
				428	340	245	1.88	1.94*
Plymouth	Fury	318	230	383	290	70	1.16	1.07
				440	350	234	1.80	1.91*

*High-performance engines probably requiring heavy-duty drive train.

9. It is puzzling that Griliches and Triplett found a significant coefficient for power in 1960, but that Triplett did not thereafter. The strong coefficients found in 1960 and 1968 suggest that Triplett's failure to find a significant value in 1961–1965 must result from his method of data selection. The insignificance of coefficients for those years might be confirmed if regressions were run on data weighted by sales, and including a large number of both six- and eight-cylinder engines.

suggests that this set of engines displays a marginal cost of engine size below that estimated in the main regression; when the weight increase is deducted, the marginal cost may be well below the estimate.

If a single value is to be taken for the marginal value of power, holding weight and number of cylinders constant, it must be one of the lower estimated values. The most probable coefficient is $1.69 per horsepower (Equation 3).

Griliches[10] has found that the presence of power steering as standard equipment causes a 30 percent rise in price. As he points out, this is not because of the value or price of power steering per se; but it reflects the fact that it is standard on only a few luxury cars, so that it captures some element of luxury omitted here. Since the treatment in this study consists of removing the cost of power steering from the price, the possibility of explanation is lost. Perhaps a more complete list of optional equipment than that reflected in the QUAL variable would improve the fit for luxury cars.

Construction of Attribute Demand Curves

THEORY

The hypothesis of Equation B.2 or B.3 above provides an equation for the marginal price of each automotive attribute, representing the value at the margin to each consumer who has purchased his equilibrium amount. In Equation B.2, this is a constant amount: so many dollars per pound, horsepower, inch, etc., at all quality levels. In Equation B.3 it is a constant percentage of the total price per unit, at all quantity levels. This is not a demand curve, but a single price for each attribute at a given point in time.

It follows that the average amount of each attribute that is purchased in the good will be a function, inter alia, of the price of that attribute as just defined. To test this hypothesis it is necessary to find markets that display different attribute prices and to compare average amounts purchased. Such markets might occur in different locations if the good is not priced nationally, or they might be found at different points in time. If other factors, such as income, that might influence the average amount purchased vary between the markets, then those factors should also be included in the test-

10. Griliches, "Hedonic Price Indexes for Automobiles."

ing equations. If tastes are the same in the different markets, so that the different prices reflect changes in the explanatory variables, then regression analysis can be used to test the hypotheses that attribute price and other explanatory variables are the primary determinants of average attribute quantity. Making the usual assumption of a constant elasticity of demand, we should be able to derive estimates of the price and income elasticity of demand for an attribute. Thus

$$\log Q^D = C + a_1 \log P + a_2 \log Y, \tag{B.4}$$

where Q is average horsepower purchased, C is a constant, P is the price of power found above, and Y is income.

One problem that arises here is errors in variables. The price observations are stochastic, since they are estimated by ordinary least squares from a set of individual prices and are accompanied by a standard error of the coefficient. Consequently, the estimated coefficients are biased and inconsistent. This problem can be mitigated by the use of an instrumental variable, the productivity of labor in the automobile industry, Z. Our price observations are from years in which the cost of producing power fell substantially, largely because of improved technology for producing higher-powered engines. Thus the productivity of labor in the automobile industry rose, and Z should be highly correlated (although negatively) with P, yet uncorrelated with the residuals in Equation B.4.

The process of solution for this instrumental variable technique consists first in regressing P by ordinary least squares on the exogenous variables Y and Z to obtain a set of fitted values for $P, \hat{P}$. Then Equation B.4 is run by ordinary least squares using the fitted values $\hat{P}$ instead of the observed values. The resulting price coefficient is the price elasticity of demand.

Actually, the model tested is a little more complex than this. It is expected that the amount of power purchased will be a lagged function of the price and income variables. This is because the automobile represents a major household expenditure, and most purchasers do not want to experiment on a good that will be with them for a few years. They tend to buy the same make of car as they previously owned[11] and to change specifi-

11. Lawrence J. White, *The Automobile Industry Since 1945* (Cambridge, Mass.: Harvard University Press, 1971).

cations, such as power, only by a small amount at each purchase. Thus it takes several years for a major price or income change to be completely reflected in the market. In addition, it takes time for manufacturers to introduce new engines, and this is done only as it becomes clear that there is a demand for them. This suggests an iterative process of introducing a larger engine, selling it, then introducing a still larger one. The time series is not long enough to permit separate estimations of the coefficients for several lagged terms of the parameters of a Koyck lag. Instead, one lag form has been assumed and tested against the unlagged model. Results are given for both unlagged variables and for this one-lag form. We do not suggest that this is the best-fitting or the true form; but it is demonstrably better than the model with no lagged variables. The form used is

$$YL_T = [0.4Y_T + 0.4Y_{T-1} + 0.2Y_{T-2}], \tag{B.5}$$

$$PL_T = [0.4P_T + 0.4P_{T-1} + 0.2P_{T-2}]. \tag{B.6}$$

DATA

Testing the demand hypothesis for an attribute of a product requires accurate data on sales, prices, and attribute quantity, and a set of markets in which different prices are believed to trace out a demand curve. We will examine the demand for automotive horsepower, using time series data beginning in 1952. Data are available for every year describing the vehicles produced in that year, including rated horsepower and price, annual sales figures by make, average power purchased, and per capita income.

We use a set of sample years rather than every year because of the expense of gathering and processing the data for each year. In a sample year, total domestic automobile sales are broken down by make and model, insofar as data permit, and then are used to select representative model observations for the study. One observation is taken for every 67,000 cars sold, so that an average annual sample includes over 50 observations, weighted by actual sales. Regressions are run for the year with price as the dependent variable and a set of physical parameters for independent variables, including weight, length, rated engine power, number of cylinders, body style, and inclusion of convenience accessories. The sample represents all domestic vehicles except that, in several years, the most expensive models, comprising up to 6 percent of total sales, do not fit well and are omitted. These regressions give prices for all parameters, including horsepower, for the sample year, expressed both in absolute terms and as a percentage of the total price.

The sample years for which regressions are run are even years from 1952 to 1960, plus 1968. The average values for major attributes and price in the sample years are shown in Table B.4. The simple correlation coefficients between these attributes and QUAL for a single year (1968) were shown earlier in Table B.2. The correlation coefficients for the marginal value of power estimated for each sample year, average power values, and real per capita income for all years from 1952 to 1968 are shown in Table B.5. The variables in Table B.5 are the basis for determining the price elasticity of demand. But they give only six points in which the price of power is known. There is a strong prior belief about the missing points, based upon the fact that changes in relative prices of different makes and models are made gradually and in small increments. For example, in the case of the 1952–1956 Ford, the reduction in marginal cost of power by a factor of three to four was accomplished with a 33 percent change in the price of the V-8 engine over four years and a big change in its power output. Other makes seem to follow a similar pattern of gradual adjustment. Thus there is a linear interpolation between the observed price of power points to fill in the time series.

It is suggested that the period of the 1950s, particularly the middle of the decade, was a time when rapid technical change greatly reduced the marginal cost of horsepower. The reduction in production cost was passed on to consumers in the form of lower marginal prices for power, leading to substantial increases in average power purchased. This was a time of relatively little net change in consumer preferences, and any change that did occur—particularly in 1957 and 1958—was nullified by 1961. The evidence for this hypothesis comes from an examination of automobile technology and the behavior of the new car market during that period.

In 1952, most domestic cars were offered with six-cylinder engines or with a straight eight. The largest manufacturer of V-8 engines was Ford, and its V-8's all used a flat head design rather than overhead valves. Compression ratios were generally below 7.5:1, and average power density was below 0.5 horsepower per cubic inch of displacement. During the next few

Table B.4. Average Vehicle Attributes in Selected Years

Attri- bute Year	Length (inches)	Weight (pounds)	Engine Dis- placement (cu in.)	Engine Power (horse- power)	Engine Compression Ratio (:1)	Price (dollars)
1952	200	3,288	237	108	6.9	2,000
1954	201	3,349	250	131	7.4	2,039
1956	203	3,416	275	181	8.2	2,244
1958	208	3,557	293	206	8.9	2,518
1960	207	3,486	268	178	8.7	2,627
1968	206	3,535	304	214		2,877
Corre- sponding Variables	LNG	WT	DIS	HP	CR	PR

From sample data. This is not a perfect representation of the data for each year because some models, particularly luxury cars and high-performance cars are not included. In general, actual averages would be slightly higher.

Table B.5. Correlation Coefficients for Regressions

	P	HP	YR
P	1	—0.778	—0.527
HP		1	+0.854
YR			1

P is price in current dollars.
HP is average power purchased in horsepower.
YR is disposable personal income in current dollars.

years, most manufacturers introduced V-8 engines with overhead valves and higher compression ratios. These and other changes permitted greater power per cubic inch of displacement and reduced engine weight per cubic inch. By 1956, compression ratios averaged almost 8.5:1, and power density was up to about 0.65 horsepower per cubic inch. These figures rose little up to 1958 and fell to the 1956 level by 1960.

The effect of increasing power per cubic inch and reducing engine weight per cubic inch was to reduce production costs per horsepower, since these costs are related primarily to the physical size of the engine and the amount of material in it. Using a V-8 instead of straight eight made these costs less nonlinear, since there seemed to be a practical upper limit to the size of in-line engines that could be built without causing structural problems in the very long crankshaft and block. The V-8, only half as long as an in-line engine of the same displacement, could withstand far greater stresses without undue wear and thus, in effect, greatly raised the practical maximum horsepower that could be produced economically. The effect was the same as a reduction in the marginal cost of power for all engines offered.

It is sometimes said that the horsepower race of the 1950s, which saw average new car horsepower rise from 104 in 1950 to 212 in 1958, reflected a change in taste of the motoring public, with an increased desire for power. If there was an increased demand for power, reflected by an upward shift in the demand curve, it should be evidenced by an increase in the marginal cost of power. Without even going to the regression equations, it is clear that this is not the case—that the marginal cost of power fell instead. For example, in 1952, the Ford V-8 cost $76 more than the six, with only 9 more horsepower and 34 pounds of extra weight in the identical body. In 1956, the price difference was up to $100, but the V-8 provided 36 more horsepower and weighed over 100 pounds more.[12] Thus a 33 percent increase in price brought four times as much added power and three times as much weight. With any reasonable allocation of values to weight, the price of power had clearly declined greatly rather than increasing. Consumers responded by purchasing more power.

If there was any change in consumer taste, other than response to changing prices, it must have been a short-term phenomenon. The horsepower

12. National Market Reports, Inc., *Red Book of Official Used Car Valuations,* October 1–November 14, 1958, Region A, Chicago.

race is said to have peaked in 1957–1958, so that, perhaps, power levels purchased in these years would be above those predicted by a model using constant tastes. By 1960, however, any particular infatuation with power per se seemed to have disappeared.

RESULTS

A series of equations was fitted to each of the sample years 1952 through 1968. The same forms were used for all years on the theory that, whatever had happened to relative prices during this period, the structure of those prices must have remained roughly the same. Table B.6 shows the results for the two best forms for all years. Looking first at Equation D, we see that between 1952 and 1954 the price of power dropped to about one-sixth of its original value, then rose and stabilized at about one-third that value. The two lowest prices, in 1954 and 1956, however, are accompanied by standard errors so large that their values must be viewed with some suspicion. Although the R^2 has remained good for cross-section data, the relationship between price and power has weakened. The same phenomenon occurs in Equation F. Here, the price drop in 1954 is only one-fifth, and the t-statistic is significant at only the 15 percent level of confidence. The equilibrium price level seems to be about one-quarter the 1952 prices for power.

One source of difference between Equations D and F is the use of weight as the vehicle size parameter in the former, and the use of length in the latter. Since the correlation coefficients between their length, weight, and power are

	WT	HP
LNG	0.91	0.53
WT	1.0	0.76

it is not surprising that exchanging length for weight should have some impact on the horsepower coefficient. Furthermore, the coefficient for number of cylinders also has a significant impact on the price of power, and is excluded in Equation D but included in F. Despite the difference in the average level of the power coefficient in the two equations, it is interesting that the movement of the two over time is parallel.

Equation F is selected as the equation that best represents theoretical expectations as to the determinants of vehicle price and that best fits the

Table B.6. Hedonic Price Results for Sample Years

Equa-tion	Vari-able	1952	1954	1956	1958	1960	1968
D	WT	0.78 (0.14)	0.50 (0.11)	0.55 (0.09)	0.23 (0.09)	0.24 (0.02)	
	HP	4.88 (1.83)	0.69 (1.26)	0.99 (0.60)	1.71 (0.45)	1.69 (0.20)	
	QUAL	70.3 (22.8)	116.3 (20.7)	72.1 (11.9)	115.1 (17.5)	126.6 (7.3)	
	R^2	0.88	0.86	0.91	0.85	0.98	
F	LNG	22.0 (5.7)	20.23 (2.77)	19.26 (1.85)	14.62 (2.97)	8.82 (0.66)	
	WT						0.51
	HP	9.79 (1.44)	2.02 (1.31)	1.61 (0.53)	2.40 (0.44)	1.91 (0.22)	2.20
	QUAL	74.78 (25.27)	108.6 (17.7)	105.2 (8.6)	115.1 (14.3)	132.2 (7.3)	47.9
	DCL	−72.51 (54.65)	22.94 (45.8)	29.36 (29.6)	−56.00 (51.1)	27.02 (20.4)	−104
	DBY	45.99 (54.15)	49.35 (26.9)	−7.26 (40.6)	−63.77 (44.7)	10.8 (29.3)	− 76.1
	R^2	0.88	0.91	0.95	0.89	0.98	0.89

WT = curb weight in pounds; HP = rated engine horsepower; QUAL = a measure of accessories and luxury; DCL = dummy variable for cylinders: = 0 in line engine; = 1 if V-8; DBY = dummy variable for body style: = 0 if 4-door sedan; = 1 if 2-door hardtop. The number in parentheses is the standard error of the coefficient.

several data sets. In large measure, this selection is based on the same criteria discussed above in the detailed analysis of the 1968 regression, and on the generally high t-statistics for the power coefficient. Because of the similar behavior of the prices in Equations D and F, it is believed that the price elasticity derived from these figures will not differ much between the two forms. Logarithmic versions of these equations also have been tested. The power coefficients follow the same general pattern here as in the linear equations. Since the latter are easier to interpret, they have been used in the analysis that follows.

To determine the price elasticity of demand for power, the price terms shown in Equation F of Table B.6 are extended by interpolation to include all years from 1952 through 1968. First, a simple single equation is run with the log of average engine horsepower as the dependent variable, and the log of the price of power and the log of disposable income as independent variables. Both price and income are in real terms. In addition, both are tried with the three-year distributed lag discussed above as Equations B.5 and B.6. The results are shown in Table B.7 as Equations 1 and 2. Note that both equations show a highly significant price coefficient, which changes by only about 10 percent when the lags are introduced. The R^2 is greatly improved by using the lagged variables, while the Durbin-Watson statistic, low in Equation 1, falls even lower in Equation 2. The residuals show a peak of power in 1957–1958 that is grossly underestimated, and an overestimate of power for 1960–1964. Equation 3 in Table B.7 shows the estimates for parameters of Equation B.4, using instrumental variables.

Replacing the price variable with the lagged price variable improves the overall fit substantially, as shown by Equation 4. The R^2 of Equation 4 is almost as great as that of Equation 2, and the t-statistic of the price term has improved over Equation 3. The Durbin-Watson statistic is low, suggesting misspecification and autocorrelation bias. The reason for this is readily apparent from a look at the residuals, which show the same great underestimate in 1957–1958 and a small overestimate in the early 1960s. Lagging the income term as well as power has no appreciable effect on either the coefficients or the explanatory power of the equations.

We now introduce the hypothesis that there was a genuine change in taste

Table B.7. Price Elasticity of Demand for Power

Equation	Method	Price Coefficient Direct	Lagged	Income Coefficient Direct	Lagged	Other	R^2	Durbin-Watson
1	OLS	−0.223 (0.057)		1.014 (0.227)			0.83	0.82
2	OLS		−0.249 (0.037)		0.841 (0.181)		0.92	0.74
3	INST	−0.187 (0.075)		1.084 (0.249)			0.83	0.70
4	INST		−0.173 (0.056)	0.982 (0.221)			0.89	0.56
5			−0.169 (0.031)	1.09 (0.13)		0.19D (0.04)	0.97	1.18

in 1957–1958, unrelated to price or income. We suppose that advertising and popular enthusiasm created, for these two years only, a special demand for power that did not exist (or existed to a lesser degree) before and after. To represent this phenomenon, we introduce the dummy variable D, which is zero in all years except 1957 and 1958, when it is one.

Equation 5 shows the dramatic effect this dummy variable has on the fit of the model. The R^2 jumps from 0.89 to 0.97 in Equation 5. Even more important, the Durbin-Watson statistic rises from 0.56 to 1.18, approaching the range of respectability. Remarkably enough, the price coefficient changes less than 3 percent despite the substantial change in fit, leaving the choice between Equations 4 and 5 unnecessary. If reliance on a two-year dummy is not thought to be justified by theory, the undummied equation gives the same answer.

Several interesting facts emerge from a study of Table B.7. First, the change in price coefficient between using ordinary least squares and instrumental variables is substantial, a drop of about 25 percent. This is a measure of the bias that would be introduced if the simultaneous nature of the system were ignored. The fact of the change itself does not prove that the instrumental formulation is the correct one. We must base our acceptance of the instrumental upon our theoretical belief that it is proper.

Within the two instrumental structures, lagging price makes a substantial improvement in the R^2, from 0.83 to 0.89, and raises the price coefficient t-statistic from 2.5 to 3.1. Adding the dummy variable for change of taste further improves the R^2 from 0.89 to 0.97 and raises the price t to 5.4. Despite these large improvements in fit, the price coefficient itself is remarkably stable, dropping only about 10 percent overall. This stability of the coefficient gives added confidence in the estimated value.

The significance of the dummy variable, combined with the always low Durbin-Watson statistic, suggests that some important determinant of the demand for automotive horsepower is not captured by price and income. Tastes and utilities are not entirely independent, and there may be periodic shifts in taste, influenced by national affairs, advertising, some mass psychology, and the like. This concession, however, does not destroy the usefulness of the theory of consumer behavior. Even without the dummy variable, Equation B.4 has substantial explanatory power. Until we can identify and incorporate in the model whatever other variables are at work here, we can still make reasonably accurate predictions of the response of consumers to changes in the price of vehicle horsepower.

The magnitude of the coefficient itself −0.173, is cause for some speculation, since this is well below the price elasticity of demand for most consumer goods. One reason, of course, would be that vehicle horsepower is not an independent good, but is a complementary good with many other vehicle attributes. The quantity of power purchased is part of a package, and represents only a small part of the total purchase price.

An independent reason is that the purchase price of power is only part of its total cost. In Appendix A, the relationship between engine power and fuel mileage was estimated. For an average vehicle, it was determined that an increase of 1 horsepower caused a decrease of 0.0217 miles per gallon. With fuel priced at $0.35 per gallon, this would cost $0.34 per year more per horsepower. If future operating costs are discounted at the 6 percent private rate, the present value of this increase in fuel cost associated with the increase in engine size would be $3.48. Thus, where we observe a change in the price of power, we should compare it, not with the magnitude of the sales price, but with the total cost of that power which is two and a half times as much. The elasticity with respect to this total cost is about twice what was estimated here, or about −0.42. This is a more respectable elasticity.

Appendix C Performance and Cost of Abatement

Uncontrolled Car (BASE)

The basis for comparison of the devices considered here is a 1963 model car with no pollution controls. The assumed uncontrolled emissions are as shown in Table C.1.

Positive Crankcase Ventilation (PCV); Change from BASE

Cost: The automobile industry estimates the cost of a closed PCV system at $12 to $15.[1] The valve, however, can be purchased at retail for $2, and hoses similar to those used for less than $1, so that it is difficult to justify a capital cost for factory installation of over $5. The valve should be replaced annually. There is no other effect on operating costs or performance, if annual replacement is performed. (See Table C.2.)

Table C.1. Emissions (Uncontrolled 1963 Vehicle Seven-Mode Cycle)

Crankcase	Evaporation	Exhaust				Total	
HC	HC	HC	CO	NO_xa	Pb b	HC	
3.15	2.77	10.2	76.9	4.0	0.133	16.12	gm/mile
—	—	834	3.3%	1,000	—	—	ppm or %

Conversions: 1.0 gm HC = 82 ppm; 23.0 gm CO = 1%; 1.0 gm NO_x = 250 ppm
Sources: For NO_x values, Paul B. Downing and Lytton Stoddard, "Benefit/Cost Analysis of Air Pollution Control Devices for Used Cars," Project Clean Air, Research Project S-10, vol. 3, University of California at Riverside, September 1970. For HC and CO values, U.S. Department of Health, Education and Welfare, Public Health Service, *Control Techniques for Carbon Monoxide, Nitrogen Oxide, and Hydrocarbon Emissions from Mobile Sources* (Washington, D.C.: U.S. Government Printing Office, 1970), Table 3–3, p. 3–8.
a. The U.S. PHS figure of NO_x of 6 gm/mile, or 1,500 ppm, is believed to be the emission rate after compression ratios rose and some devices were added to control exhaust hydrocarbons; the Downing figure seems more reasonable for a precontrol car. This is confirmed in tests which General Motors ran before and after installing emission control kits on used cars: 101 used cars, built in 1954 through 1966 and averaging an age of seven years in 1970, produced precontrol emissions of 626 ppm hydrocarbons, 3.19 percent carbon monoxide, and 977 ppm oxides of nitrogen in hot cycles of the 1970 Federal Test Cycle. (See G. W. Niepoth, G.P. Ransom, and J. H. Currie, "Exhaust Emission Control for Used Cars," Paper no. 710069, Society of Automotive Engineers, Automotive Engineering Congress, January 1971.) Use of the PHS NO_x figure would entirely mask the NO_x increases caused by the early HC and CO controls.
b. Pb is from 2.5 grams per gallon of gasoline, 15 miles per gallon, and an 80 percent emission rate.

1. U.S. Department of Health, Education and Welfare, Public Health Service, *Control Techniques for Carbon Monoxide, Nitrogen Oxide, and Hydrocarbon Emissions from Mobile Sources* (Washington, D.C.: U.S. Government Printing Office, 1970), p. 5–22.

Effect: The closed positive crankcase ventilation system entirely elimi-
nates crankcase emissions. These average 3.15 gm/mile of hydrocarbons.
Total hydrocarbons are thus reduced by 20 percent. Other emissions are not
significantly affected. (See Table C.3.) Estimates for the five indices described in Chapter 3 and the associated percentage costs compared with
BASE are as shown in Table C.4.

Comment: Since several pollutants were not affected by the PCV systems,
indicators 3 and 6 have zero percentage change. Since the cost of the device
is positive, the percentage cost for these indicators is infinite. This means
that these indicators cannot be used to evaluate this device, and some others
must be relied upon. Indicators 2 and 5 are suggested as those best reflecting

Table C.2. Cost (over BASE)

Purchase ($)	Power loss ($)	Total Capital ($/year)	Maintenance ($/mile)	Fuel ($/mile)	Total Variable ($/mile)	Total ($/mile)
5	—	0.680	0.0002	—	0.0002	0.00027

Table C.3. Effect (Seven-Mode Cycle)

			Total				
	Crankcase	Evaporation	Exhaust				
Pollutant	HC	HC	HC	CO	NO_x	Pb	HC
Gross (gm/mile)	0	2.77	10.2	76.9	4.0	0.133	13.0
Change from BASE (gm/mile)	−3.15	—	—	—	—	—	− 3.15
Change from BASE (% of BASE)	−100	0	0	0	0	0	−20

Table C.4. Effect on Indices (compared with BASE)

Index	2 Net Percent	3 Minmax Percent	4 Percent Reduced	5 (2) without Lead	6 (3) without Lead
Percentage Change (%)	−5	0	−20	−6.7	0
$\dfrac{\$/mile}{Percent} \times 10^4$	0.54	∞	0.13 HC	0.40	∞

the uncertainty about the effects of the pollutants; they can be compared among all devices evaluated here, since they are always finite.

1968 Clean Air Package (68 CAP); Change from PCV

Cost: The automobile industry estimates the capital cost of the 68 CAP at $18 to $25. The Bureau of Labor Statistics, however, estimates it at only $16[2] and the Environmental Protection Agency (EPA) at $2; because relatively few new components are actually added, the BLS estimate is probably reliable. There are no figures to suggest that maintenance costs rise with the addition of the 68 CAP, since much of it consists of changes in existing components. Gas mileage is not affected, except that the leaner adjustment may improve fuel economy on some engines. Peak power is unchanged. Since they reflect an overall cost reduction, the EPA costs also are presented.[3] (See Table C.5.)

Effect: The 68 CAP greatly reduces both hydrocarbons and carbon monoxide but also raises the emission of oxides of nitrogen by raising peak combustion temperatures. Prototypes submitted for federal testing all met the 1968 federal standard of 275 ppm of exhaust hydrocarbons and 1.5 percent of exhaust carbon monoxide. Later sampling of actual highway vehicles, however, showed exhaust emissions of 325 ppm and 1.55 percent, respectively; oxides of nitrogen rose markedly from the PCV vehicle.[4] (See Table C.6.) Estimates for the five indices and the associated percentage costs compared with the PCV vehicle are shown in Table C.6.

Table C.5. Cost (over PCV)

Purchase ($)	Power Loss ($)	Total Capital ($/year)	Mainte- nance ($/mile)	Fuel ($/mile)	Total Variable ($/mile)	Total ($/mile)
16	—	2.176	—	—	—	0.00022
2 (EPA)	—	0.272	—	(—0.0005)	(—0.0005)	(—0.00047)

2. Ibid.
3. Environmental Protection Agency, *The Economics of Clean Air,* Report of the Administrator to Congress, Senate Doc. no. 92–6 (Washington, D.C.: U.S. Government Printing Office, March 1971), p. 3–16.
4. A. J. Hocker, "Exhaust Emissions from Privately Owned 1966–1969 California Automobiles: A Statistical Evaluation of Surveillance Data," California Air Resources Laboratory, Los Angeles, Supplement to Progress Report no. 16, November 7, 1969, p. 7.

Table C.6. Effect

Pollutant	Crank-case HC	Evapo-ration HC	Exhaust HC	CO	NO$_x$	Pb	Total HC
Gross (gm/mile)	0	2.77	4.38	38.3	6.08	0.133	7.15
Change from PCV (gm/mile)	0	—	−5.82	−38.6	2.08	—	5.85
Change from PCV (% of BASE)	0	—	−57	−50	52	-—	36

Table C.7. Effect on Indices (compared with PCV)

Index	2 Net Percent	3 Minmax Percent	4 Percent Reduced	5 (2) with-out Lead	6 (3) with-out Lead
Percentage Change (%)	−8.5	52	−86	−11.3	52
$\dfrac{\$/mile}{Percent} \times 10^4$	0.25	∞	0.025	0.19	∞

Controlled Combustion System (70 CCS); Change from PCV

Cost: The automobile industry has estimated the cost of engine modification to meet the 1970 U.S. standards at $36 to $45 per vehicle. The Bureau of Labor Statistics, however, estimates the cost at only $21.50 [5] and the Environmental Protection Agency at $7.[6] The BLS and EPA estimates seem reasonable, given the small amount of added equipment; but to be conservative, $25 will be used as the increase in cost over a PCV vehicle. Again, since there is no additional maintenance specified for the 70 CCS system, operating cost is assumed unchanged. Gas mileage is not affected, and peak power is not reduced in most engines. (See Table C.8.)

Effect: The 70 CCS further reduces hydrocarbons and carbon monoxide, and again raises the oxides of nitrogen, since it is primarily an extension of the 68 CAP. Prototypes submitted for federal testing all met the 1970 standards of 2.2 gm/mile of hydrocarbons and 23 gm/mile of carbon monoxide.

5. U.S. Department of Health, Education and Welfare, Public Health Service, *Control Techniques*, p. 5–22.
6. Environmental Protection Agency, *Economics of Clean Air*, p. 3–16.

Table C.8. Cost (over PCV)

Purchase ($)	Power Loss ($)	Total Capital ($/year)	Mainte-nance ($/mile)	Fuel ($/mile)	Total Variable ($/mile)	Total ($/mile)
25	—	3.40	—	—	—	0.00034
7 (EPA)	—	0.95	—	(—0.0005)	(—0.0005)	(—0.00041)

Table C.9. Effects

				Total				
	Crank-case	Evapo-ration	Exhaust					
Pollutant	HC	HC	HC	CO	NO$_x$	Pb	HC
Gross (gm/mile)	0	2.77	3.45	25.0	7.0	0.133	6.22
Change from 1968 (gm/mile)	0	0	—0.93	—13.3	0.92	0	0.93
Change from PCV (gm/mile)	0	0	—6.75	—51.9	3.0	0	6.78
Change from PCV (% of BASE)	0	0	—66	—67	75	0	—42

Table C.10. Effect on Indices (compared with PCV)

Index	2 Net Percent	3 Minmax Percent	4 Percent Reduced	5 (2) with-out Lead	6 (3) with-out Lead
Percentage Change (%)	—8.5	75	—109	—11.3	75
$\dfrac{\text{\$/mile}}{\text{Percent}} \times 10^4$	0.40	∞	0.03	0.30	∞

In the past, however, lifetime average emissions have been well above the emissions of the prototypes and well above the standards. The following data are based upon extrapolation of previous relationships between proto-type performance and production vehicle performance. (See Table C.9.) From this, it is clear that abatement of hydrocarbons and carbon monoxide has advanced over the 68 CAP vehicle, but emissions of oxides of nitrogen again are seriously increased. Estimates for the five indices and the associ-ated percentage costs compared with the PCV vehicle are given in Table C.10. Note that the average cost of abatement rose about 50 percent in 1970 over 1968 for all measures except number 4, which relates the cost only to those pollutants which are actually reduced. Unless only hydrocarbons and carbon monoxide are of concern, abatement costs are rising rapidly.

Low-Lead, Low-Octane Engine (71 LLO); Change from 70 CCS

Cost: The only change that affects manufacturing costs is the transmission-controlled spark advance equipment. Because of its relative simplicity, its cost is estimated at $10. In addition, there is a decrease averaging 5 percent in power for engines of a given displacement, resulting from the lower com-pression ratios. This is the same as raising the average and marginal cost of power by 5 percent, or from $1.69 to $1.775 per horsepower. With a price elasticity of demand for power of -0.173, this means that 0.85 percent less power will be purchased. The cost of these changes is

$$C = \tfrac{1}{2}(P_2 - P_1)(Q_1 + Q_2) = \tfrac{1}{2}(0.085)(230 + 228.045)$$
$$= 19.47.$$

Another cost increase comes from the reduction in fuel economy, a direct consequence of lower compression ratios. In the range of 8.5 to 10.5:1, the relationship between gas mileage and fuel economy can be expressed by

$$\text{MPG} = k \times \frac{\text{CR}}{1 + .0625\text{CR}},$$

where k is a constant and CR is the compression ratio.[7] Thus reducing the

7. See Appendix A.

compression ratio from 9.5 to 8.5 should increase fuel consumption by 4
percent. The other changes in ignition and valve timing have been said to
cause a total reduction of 10 percent in gas mileage,[8] but a more conservative figure of 6 percent will be used here. This reduces gas mileage from 15
to 14.1 MPG, which at $0.35 per gallon for gas costs $0.0014 per mile. The
resource cost, based on $0.24 per gallon, is $0.0010 per mile.

If lead pollution were not a matter of concern, one could make the
changes described above, burn lower than regular octane fuel (91 instead of
94 Research Octane Number or RON), and save money on the price of the
fuel, perhaps $0.01 per gallon. Thus all the above costs could be attributed
to lowering oxides of nitrogen. The use of lower octane fuel, however, provides an ideal opportunity to lower the lead content of the gasoline, since
lead is used to provide higher octane. There has been substantial dispute
over the cost of providing low-lead or no-lead fuels, but one alternative can
be evaluated with some certainty. Present regular gas is 94 RON with 2.3
grams of lead per gallon, and premium is 97+ RON with 2.8 grams of lead
per gallon. Estimates of the cost of selling a 91 RON gas with 0.5 grams of
lead per gallon, and of the 94 and 97 RON, range from an addition of $0.01
per gallon to a saving of $0.01 per gallon for current three-pump stations.[9]
It seems reasonable to assume that the octane reduction offsets the reduced
lead use, so that the 91 RON low lead will cost the same as 94 RON with
2.5 grams of lead per gallon. A leaded (2.5 grams) 91 RON gas probably
would be even cheaper, but no recent estimates have been made of how
much cheaper. Thus we will have to attribute the cost of this plan equally
to reductions of lead and oxides of nitrogen.

Maintenance costs for this automobile are unchanged. There may be
benefits to the engine, spark plugs, and exhaust system from the presence
of less lead in the fuel, but no reliable data are available on this. Thus, the
cost cited in Table C.11 probably overstates the real cost of this plan.

8. *Oil and Gas Journal*, February 23, 1970, p. 59.
9. The president of Union Oil Company has said that this grade would cost about a penny
a gallon more than regular (quoted in *Oil and Gas Journal*, March 9, 1970, p. 28). GM's
president, Edward N. Cole, said that 91 RON regular with 0.5 gm Pb/gal with a leaded
97 RON premium could be sold for about current prices (quoted in *Oil and Gas Journal*,
March 16, 1970, p. 92). Gulf Oil Corporation is currently marketing a low-octane, low-lead
gas (0.5 gm Pb/gal) for one cent below regular (*Boston Globe* advertisement, Friday,
August 14, 1970, p. 9).

Table C.11. Cost (over 70 CCS)

Type of Cost	Pur- chase ($)	Power Loss ($)	Total Capital ($/year)	Mainte- nance ($/mile)	Fuel ($/mile)	Total Variable ($/mile)	Total ($/mile)
User	10	19.47	4.00	0	0.0014	0.0014	0.00180
Resource	10	19.47	4.00	0	0.0010	0.0010 Over PCV	0.00140 0.00174
EPA	17a (over PCV)		2.31			(—0.00027)	(—0.00004)

a. Includes TANK.

Table C.12. Effect

			Total				
	Crank- case	Evapo- ration	Exhaust				
Pollutant	HC	HC	HC	CO	NO$_x$	Pb	HC
Gross (gm/mile)	0	2.77	3.45	25.0	4.5	0.027	6.22
Change from 1970 (gm/mile)	0	0	0	0	—2.5	—0.106	0
Change from PCV (gm/mile)	0	0	—6.75	—51.9	0.5	—0.106	—6.78
Change from 1970 (% of BASE)	0	0	0	0	--63	—80	0
Change from PCV (% of BASE)	0	0	—66	—67	12	—80	—42

Effect: The results must be projected from tests of prototypes and from theoretical considerations. It is assumed that carbon monoxide is un- changed. While unleaded gasoline reduces hydrocarbons by reducing engine deposits, similar benefits for low-lead gas are short-lived and do not significantly affect lifetime emissions. Oxides of nitrogen are assumed to be reduced by about 36 percent, a conservative estimate for the many changes here. This would leave them about 12 percent above 1971 California stan- dards, an excess typical of previous hydrocarbon emissions over specified standards. Lead is reduced by 80 percent, the percentage reduction in fuel lead content. (See Table C.12.) This table shows that the primary result of this plan is a reduction in lead emissions and a secondary result is a reduc- tion in emissions of oxides of nitrogen. Estimates of the five indices compared with the 70 CCS and with PCV and associated resource percentage costs are shown in Table C.13.

Evaporation Emission Control (TANK); Change from Any Vehicle

Cost: The automobile industry estimates the cost of this evaporation control at $36 to $50 per vehicle.[10] For the systems described above, independent estimates have been about two-thirds of the lower industry estimate, and it has been suggested that production costs are well below those first encountered. Here, since we have no separate estimate, a compromise will be made between the lower estimate and two-thirds thereof, or $24. The only maintenance required is replacement of the purge air filter, for which $2 per year is allowed. Fuel mileage is unaffected or slightly improved. Performance is unaffected. (See Table C.14.)

Effect: The 1971 federal standard for evaporation calls for no more than

Table C.13. Effect on Indices (compared with 70 CCS and PCV)

Index	2 Net Percent	3 Minmax Percent	4 Percent Reduced	5 (2) without Lead	6 (3) without Lead
From 70 CCS: Percentage Change (%)	−35.75	0	−143	−21	0
$/mile ──── × 10⁴ Percent	0.39	∞	0.098	0.67	∞
From PCV: Percentage Change (%)	−44	12	−189	−32	12
$/mile ──── × 10⁴ Percent	0.40	∞	0.09	0.54	1.45

Table C.14. Cost (over any vehicle)

Purchase ($)	Power Loss ($)	Total Capital ($/year)	Mainte- nance ($/mile)	Fuel ($/mile)	Total Variable ($/mile)	Total ($/mile)
30	0	4.08	0.0002	0	0.0002	0.00061

10. U.S. Department of Health, Education and Welfare, Public Health Service, *Control Techniques,* p. 5–22.

6 grams of hydrocarbons emitted per test, or 0.49 grams per mile on the average. It will be assumed here that this standard is met. Since the absorbed hydrocarbons are ultimately drawn through the engine, they may affect exhaust emissions. It is assumed that the charcoal is purged at a sufficiently slow rate that exhaust emissions are not significantly increased. This raises an important technical question: If automobile engines rarely run with excess air (a fuel shortage), then is it not inevitable that all extra hydrocarbons that enter the carburetor will leave the tailpipe unburned, assuming no exhaust system antipollution device? Depending on the answer to this question, our present results could be changed. (See Table C.15.) Estimates for the five indices compared with any car are given in Table C.16.

Exhaust Gas Recirculation (EXR); Change from 71 LLO

Cost: No production figures are available for the cost of installing exhaust recirculation on new cars. The simple design tested by General Motors has

Table C.15. Effect

Pollutant	Crank-case HC	Evapo-ration HC	Exhaust HC	CO	NO_x	Pb	Total HC
Gross (gm/mile)	0	0.49	Not Affected				
Change from Any Vehicle (gm/mile)	0	−2.28	0	0	0	0	−2.28
Change from Any Vehicle (% of BASE)	0	−82	0	0	0	0	−14

Table C.16. Effect on Indices (compared with any vehicle)

Index	2 Net Percent	3 Minmax Percent	4 Percent Reduced	5 (2) with-out Lead	6 (3) with-out Lead
Percentage Change (%)	−3.5	0	−14	−4.7	0
$\dfrac{\$/mile}{Percent} \times 10^4$	1.75	∞	0.44 HC	1.30	∞

no moving parts and appears to involve minimal installation cost.[11] Twenty-five dollars should be quite generous for all capital costs. In addition, the use of 15 percent recirculation cuts maximum power by 15 percent, or raises the average and marginal cost of power by that amount. This introduces a cost of $60.10 per vehicle for power loss. As a result of using no lead, the cost of gasoline is raised by $0.0045 per gallon (see 71 CAT). Finally, gasoline mileage is reduced by the system, in the amount of about 4 percent.[12] This raises fuel consumption from 14.1 MPG to 13.5 MPG, at a cost, combined with the price increase to $0.3545 per gallon for gas of $0.0014 per mile, and a resource cost of $0.0011 per mile. There appears to be no added maintenance for this system if unleaded fuel is used. The EPA costs are for a recirculation system sufficient to meet 1973 U.S. standards. (See Table C.17.)

Effect: Oxides of nitrogen are reduced by 54 percent. With both the exhaust recirculation and unleaded fuel, there is no significant change in hydrocarbons or carbon monoxide. The effects are summarized below. There is a risk that owners will disconnect or plug the system, to regain peak power and better gasoline mileage. (See Tables C.18 and C.19.)

Catalytic Exhaust Converter with Unleaded Gas (71 CAT); Change from EXR

Cost: Because the catalytic converter is not now a commercial product, no production cost figures are available. The EPA estimates are that new

Table C.17. Cost (over 71 LLO)

Purchase ($)	Power Loss ($)	Total Capital ($/year)	Mainte-nance ($/mile)	Fuel ($/mile)	Total Variable ($/mile)	Total ($/mile)
25	60.10	9.65	0	0.0011	0.0011	0.00206
					Over PCV	0.00380
42 a	EPA	5.71	0.00104	(—0.00025)	0.00079	0.00136

a. Over PCV.

11. G. S. Musser, J. A. Wilson, R. G. Hyland, and H. A. Ashby, "Effectiveness of Exhaust Gas Recirculation with Extended Use," Paper no. 710013, Society of Automotive Engineers, Automotive Engineering Congress, January 11–15, 1971.
12. W. Glass, F. R. Russell, D. T. Wade, and D. M. Hollobaugh, "Evaluation of Exhaust Recirculation for Control of Nitrogen Oxides Emission," Paper no. 700146, Society of Automotive Engineers, Automotive Engineering Congress, January 1970.

Table C.18. Effect

Pollutant	Crank-case HC	Evapo-ration HC	Exhaust HC	CO	NO$_x$	Pb	Total HC
Gross (gm/mile)	0	2.77	3.45	25.0	2.07	0	6.22
Change from 71 LLO (gm/mile)	0	0	0	0	2.43	−0.027	0
Change from 71 LLO (% of BASE)	0	0	0	0	−61	−20	0
Change from PCV (gm/mile)	0	0	−6.75	−51.9	−1.93	−0.133	−6.75
Change from PCV (% of BASE)	0	0	−66	−67	−48	−100	−42.1

Table C.19. Effect on Indices (compared with 71 LLO)

Index	2 Net Percent	3 Minmax Percent	4 Percent Reduced	5 (2) without Lead	6 (3) without Lead
From LLO:					
Percentage Change (%)	−20.3	0	−18	20.3	0
$\dfrac{\$/\text{mile}}{\text{Percent}} \times 10^4$	1.01	∞	0.25	1.01	∞
From PCV:					
Percentage Change (%)	−64.3	−42	−257	−52.3	−42
$\dfrac{\$/\text{mile}}{\text{Percent}} \times 10^4$	0.57	0.88	0.14	0.71	0.88
$\dfrac{\$/\text{mile}}{\text{Percent}} \times 10^4$	0.21				EPA

car installation may cost $198 per vehicle above EXR costs, or $240 above uncontrolled vehicle costs.[13] It is assumed that the useful life of the catalyst will be 50,000 miles on unleaded fuel. The replacement cost is assumed to be an additional $98, but a $50 credit is given for normal exhaust system maintenance. Rated power is assumed to be unaffected, although, in practice, the increased exhaust back pressure should reduce it. Gas mileage is reduced 5 percent.

The cost of fuel is increased, since the same octane (91 RON) must be produced without lead. This increase is taken here at $0.0045 per gallon.[14] This will increase operating cost, at 14.1 MPG, by $0.00032 per mile. (See Table C.20.)

Effect: After the pollutant gases leave the engine, they are converted to harmless gases including carbon dioxide, nitrogen, and water. Emission rates are assumed to be the 1975 standards plus 25 percent, although there is currently serious doubt that this performance can be achieved. (See Tables C.21 and C.22.)

Table C.20. Cost (over EXR)

Purchase ($)	Power Loss ($)	Total Capital ($/year)	Mainte- nance ($/mile)	Fuel ($/mile)	Total Variable ($/mile)	Total ($/mile)
198	0	26.93	0.0007	0.00169	0.00239	0.00508
					Over PCV	0.00888
					EPA over PCV	0.00644

13. EPA, *Economics of Clean Air,* p. 3–16. This is an estimate for meeting the 1975 standards. The Ford Motor Company estimated that capital costs for 1975 would be from $290 to $500 more than the cost for 1973, while maintenance cost would increase by $25 over 50,000 miles, with no catalyst replacement. (Application of the Ford Motor Company to the Environmental Protection Agency, October 13, 1972, pp. 10, 102.) This produces marginal costs of $0.0044 to $0.0073 per mile, compared to $0.00508 in Table C.20. Another estimate made by the government placed the costs substantially lower with a capital cost of $214.00. Committee on Motor Vehicle Emissions, Division of Engineering, *Semiannual Report* to the Environmental Protection Agency (Washington, D.C.: National Academy of Sciences, January 1, 1972), p. 43.
14. The Panel on Automotive Fuels and Air Pollution, *The Economics of Unleaded Gasoline,* November 1970, Appendix, p. 29, Table IV, 5. The national average increase in gas cost for a 91 RON unleaded, with 94 and 100 RON leaded fuels, is $0.0018 per gallon. If all of this is attributed to the leaded grade, it is $0.0045 per gallon.

Table C.21. Effect [a]

	Crank-case HC	Evapo-ration HC	Total				
			Exhaust				Total
Pollutant			HC	CO	NO$_x$	Pb	HC
Gross (gm/mile)	0	2.77	0.69	2.3	2.07	0	3.46
Change from EXR (gm/mile)	0	0	−2.76	−22.7	0	−0.027	−2.76
Change from EXR (% of BASE)	0	0	−27	−29.5	0	−20	−17
Change from PCV (% of BASE)	0	0	−93	−97	−48	−100	−78.5

a. The 1975 standards are CO 0.34 gm/mile, HC 0.41 gm/mile (exhaust), and NO$_x$ 3.0 gm/mile. These are CVS, and are increased 25%, then converted to 7-mode above, except that NO$_x$ is assumed to be further reduced in preparation for 1976 standards.

Table C.22. Effect on Indices (compared with EXR or PCV vehicle)

Index	2 Net Percent	3 Minmax Percent	4 Percent Reduced	5 (2) without Lead	6 (3) without Lead
From LLO: Percentage Change (%)	−16.6	0	− 66.5	−16.5	0
$\dfrac{\$/\text{mile}}{\text{Percent}} \times 10^4$	3.06	00	0.76	3.08	00
From PCV: Percentage Change (%)	80.9	−48	−323.5	−74.5	−48.0
$\dfrac{\$/\text{mile}}{\text{Percent}} \times 10^4$	1.10	1.85	0.24	1.19	1.85

Comment: The costs and performance reported are for technology that is not yet fully developed. The capital and maintenance costs could be seriously in error. The results here, then, are not a final point for judging catalysts, but a reference point for evaluating them *if* they can be produced for this cost and performance standard.

Some low-cost means of ensuring that the catalyst is replaced is necessary. The easiest system would be to require that all five-year-old cars obtain a certificate from a mechanic stating that the catalyst had been replaced in the last year. This would be a prerequisite to registration in the sixth year. In states where there is an annual safety inspection, the mechanic could be required to find a new catalyst in all automobiles driven over 50,000 miles.

For adequate performance, it is essential that leaded gasoline never be used. The effectiveness of the catalyst can be significantly reduced by even a single tank of leaded gasoline. Thus, for full effectiveness, it will be necessary to prohibit sales of all leaded gasoline, or to design gas tank openings and filler nozzles so that the nozzle of a pump with leaded gas cannot be inserted in the tank of a CAT car.

If it is impossible to use leaded gas, and if replacement of the catalyst is required at the appropriate time, this system should be as effective as the figures show. If not, it may not have much impact after a few years. It is deemed less impervious to owner neglect than devices currently in use.

Gaseous Fuel Conversion (71 CNG); Change from 71 LLO with TANK

Cost: Installation costs are estimated at \$350 per vehicle.[15] Peak engine power is cut by about 15 percent when operating on CNG.[16] This is the same as a 15 percent rise in the marginal and average cost of power, and given a price elasticity of demand for power of 0.17, this would reduce power by 2.5 percent. Starting with 228.045 HP and \$1.78 per HP (from 71 LLO) the cost of these changes would be as follows:

$$C = \tfrac{1}{2}(P_2 - P_1)(Q_1 + Q_2)$$
$$= \tfrac{1}{2}(2.05 - 1.78)(228.045 + 222.230)$$
$$= \$60.10.$$

15. *Federal Low-Emission Vehicle Procurement Act,* Joint Hearings before the Subcommittee on Energy, Natural Resources and the Environment of the Committee on Commerce, and the Subcommittee on Air and Water Pollution of the Committee on Public Works, U.S. Senate, 91st Congress, January 27–29, 1970, p. 36.
16. Ibid., p. 183.

Usually the heat energy of 100 cubic feet (cf) of natural gas is about 13 percent less than that of a gallon of gasoline. In stop-and-go driving, however, natural gas effects some savings, so that about the same mileage is obtained from 100 cf of gas and one gallon of gasoline.[17] The domestic space heating price for natural gas in Cambridge, Massachusetts, is $0.1123 per 100 cf, at the highest consumption rate (lowest marginal cost) in 1971. Increasing this by the ratio between gasoline retail and dealers' prices, to allow for compressing and filling, gives about $0.15 per 100 cf.[18] This compares with a current retail gasoline price of $0.35 per gallon with tax and $0.24 without. Thus, there is a saving of $0.20 per gallon in user cost, if the CNG is untaxed, and $0.09 per gallon in resource costs: $0.0142 and $0.0064 per mile.

Engine maintenance is said to be reduced because of the absence of lead and the inability of CNG to gum up the engine with deposits. Since no figures are available, this cannot be quantified here. Assuming 10,000 miles per year operation, as before, the costs are as shown in Table C. 23.

The second set of costs represents those for vehicles driven 25,000 miles per year and lasting five years; this may be more typical for intensively used fleets.

Here we have the remarkable appearance of a control scheme that saves money, even if the fuel is fully taxed. Why is it not universally adopted? First, because the range is short, most people could operate on CNG only part of the time and would thus have smaller savings. Second, because the fueling stations do not exist, generally, the owner of a converted car cannot refuel conveniently. The system becomes practicable only where the vehicles

17. Ibid., p. 184.
18. The comparison with gasoline price ratios reflects the similarity in labor input to the filling operation whether the fuel is gasoline or natural gas. In either case, an attendant must be present to connect the fuel source to the vehicle, record the quantity pumped, and disconnect the fuel source. The major difference is that natural gas requires no large underground storage tanks, but does need a high-pressure compressor and a small high-pressure compressor storage tank. It is not intuitively obvious how these differences affect costs, but in California, where residential natural gas costs $0.10 per 100 cf, CNG delivered to fleet cars is only $0.13 per 100 cf. Thus the $0.15 per 100 cf figure is probably high. Much higher prices are quoted for filling and exchanging CNG tanks in some areas, but these do not reflect the economies of a large volume operation, which would be associated with fleet use. These prices are for 1971, and if the price of natural gas relative to gasoline continues to rise as it has since 1971, the cost of this device will increase correspondingly.

Table C.23. Cost (over 71 LLO) [a]

Type of Cost	Pur- chase ($)	Power loss ($)	Total Capital ($/year)	Mainte- nance ($/mile)	Fuel ($/mile)	Total Variable ($/mile)	Total ($/mile)
User[b]	350	60.10	55.77	less	−0.0142	−0.0142	−0.0086
Resource[b]	350	60.10	55.77	less	−0.0064	−0.0064	−0.0008
						Over PCV	+0.00055
User[c]	350	60.10	97.32	less	−0.0142	−0.0142	−0.0103
Resource[c]	350	60.10	97.32	less	−0.0064	−0.0064	−0.0025

a. Negative entries indicate a cost *saving*.
b. 10,000 mi/year, 10-year life.
c. 25,000 mi/year, 5-year life.

return frequently to a central garage where the owner can invest in a fueling station to assure his own supply. Finally, a substantial amount of trunk space is lost; and because many individuals would find this inconvenient, like lost power, it must involve some imputed cost.

Effect: The available data on CNG emissions pertain to pre-1970 converted vehicles.[19] The emission from a converted 71 LLO car should be even lower because of the lower compression ratio and other changes. Also an important new factor is introduced here: in addition to reducing HC emissions, the CNG fuel produces a different composition of HC, which is far less reactive in producing photochemical smog than the HC from gasoline fuels. The measurements below will use the HC adjusted downward for reactivity; the unadjusted figure is 1.9 gm/mile. The change is compared with both 1971 LLO and the PCV car, to get the marginal cost and average difference. It is assumed that all mileage is on the CNG fuel, but the gas tank is filled. (See Table C.24.)

The table shows the very complete reduction in all pollutants from base levels: all are less than 20 percent of base emissions, including the PCV and evaporation controls. Since there is a cost saving over 71 LLO, it hardly makes sense to compute the cost per dollar of abatement on that basis: it is zero or negative. Instead, the cost and performance of 71 LLO, TANK, and CNG together will be computed, with total cost of $0.00155. The 10,000 mile data will be used here, with resource costs.

19. *Federal Low-Emission Procurement Act.*

Table C.24. Effect (over 71 LLO + TANK and PCV)

Pollutant	Crank-case HC	Evapo-ration HC	Total Exhaust HC	CO	NOx	Pb	HC
Gross (gm/mile)	0	2.77	0.6	7.0	0.6	0	3.37
Change from LLO + TANK (gm/mile)	0	0	−2.85	−18.0	−3.9	−0.027	−2.85
Change from PCV (gm/mile)	0	0	−9.63	−69.9	−3.4	−0.133	−9.63
Change from PCV (% of BASE)	0	0	−94	−91	−85	−100	−60

Table C.25. Effect on Indices (compared with PCV)

Index	2 Net Percent	3 Minmax Percent	4 Percent Reduced	5 (2) without Lead	6 (3) without Lead
Percentage Change (%)	−84.0	−60	−336	−78.7	−60
$\dfrac{\$/mile}{Percent} \times 10^4$	0.18	0.26	0.05	0.20	0.26

Comment: It must be remembered that the limited range and the trunk space reduction make this applicable to only a limited number of vehicles without substantial added cost. It cannot be compared on an equal basis with other abatement systems. The system is perfectly reliable, however, in that no special maintenance is required to continue the rated emission levels, as is the case with other advanced pollution control systems. Thus, it is a certain means of reducing emissions, where it can be used.

Modified Gasoline Composition (GCOMP); Change from Typical Gasoline Composition, 71 LLO

Cost: It has been estimated that reducing the RVP of gasoline from 8.6 to 6.0 would cost large refiners $0.0133 per gallon. Removing light olefins up through C_5 would cost $0.0104 per gallon. With gasoline mileage of 14.1 MPG, these costs amount to $0.00095 and $0.00074 per mile.[20]

Effect: Reducing the RVP cuts gross evaporative hydrocarbons by about 50 percent and evaporative reactivity by 60 percent. There is a small increase in exhaust emissions, so that the net result is a 50 percent reduction in reactive emissions. Removing the light olefins does not significantly affect gross hydrocarbon emissions, but does reduce reactive evaporative emissions by 60 percent, with little impact on exhaust emissions.[21] None of these changes seems to have a significant and consistent impact on other emissions. Because light olefin removal costs less and does more than RVP reduction, it alone will be evaluated. (See Table C.26.)

Estimates for the five indices compared with any car are given in Table C. 27. A comparison of this with the evaporative controls of TANK shows that it is less effective and costs significantly more per centage reduced, even when hydrocarbons are measured on a reactivity basis, as they are here. Once mechanical evaporative controls are installed, the effectiveness of gasoline composition modifications is much less because there is a smaller emission base to work on.

Table C.26. Effect

Pollutant	Crank-case HC	Evapo-ration HC	Exhaust HC	CO	NO_x	Pb	Total HC
Gross (gm/mile)	0	1.11	———————— Not Affected ————————				
Change from Any Vehicle (gm/mile)	0	—1.66	———————— Not Affected ———————				—1.66
Change from Any Vehicle (% of BASE)	0	—60	——————— Not Affected ———————				—10.3

20. James W. Daily, "Los Angeles Gasoline Modification: Its Potential as an Air Pollution Control Measure," *Journal of the Air Pollution Control Association,* 21, no. 2 (February 1971):70–80.
21. Basil Dimitriades, B. H. Eccleston, and R. W. Hurn, "An Evaluation of the Fuel Factor Through Direct Measurement of Photochemical Reactivity of Emissions," *Journal of the Air Pollution Control Association,* 20, no. 3 (March 1970):150–160.

Table C.27. Effect on Indices (compared with any car)

Index	2 Net Percent	3 Minmax Percent	4 Percent Reduced	5 (2) with- out Lead	6 (3) with- out Lead
Percentage Change (%)	−2.58	0	−10.3	3.43	0
$\dfrac{\$/\text{mile}}{\text{Percent}} \times 10^4$	2.86	∞	0.72	2.15	∞

Table C.28. Cost (on any 1966–1968 vehicle)

Purchase ($)	Power Loss ($)	Total Capital ($/year)	Mainte- nance ($/mile)	Fuel ($/mile)	Total Variable ($/mile)	Total ($/mile)
0	0	0	0.0003	0	0.0003	0.0003

Table C.29. Effect (on 68 CAP and PCV)

Pollutant	Crank- case HC	Evapo- ration HC	Exhaust HC	CO	NO_x	Pb	Total HC
Gross (gm/mile)	0	2.77	3.95	32.2	6.44	0.133	6.72
Change from 68 CAP (gm/mile)	0	0	−0.43	− 6.1	0.36	0	−0.43
Change from 68 CAP (% of BASE)	0	0	−4.2	− 7.9	9.0	0	−2.7

Minor Engine Tune-up (TUNE); Change from a Typical 66-68 CAP or Noncontrolled Used Vehicle

Cost: There is, of course, no capital cost. The cost of performing the tune-up has been given as $3,[22] which can be accepted only if we ignore the owner's cost in getting the vehicle to the garage, leaving it, or waiting for it. We must therefore assume that the tune-up is added to regular maintenance or an inspection, either safety or pollution. Because only idle parameters are affected, fuel economy is not changed significantly, nor are other maintenance costs or power outputs. (See Tables C.28 and C.29.)

22. *Federal Low Emission Procurement Act.*

Table C.30. Effect on Indices (compared with 68 CAP)

Index	2 Net Percent	3 Minmax Percent	4 Percent Reduced	5 (2) with- out Lead	6 (3) with- out Lead
Percentage Change (%)	−0.40	9.0	10.6	−0.53	9.0
$\dfrac{\$/mile}{Percent} \times 10^4$	7.5	∞	0.28	5.66	∞

It should be noted that the tune-up is similar to the original 68 CAP in that reduction in hydrocarbons and carbon monoxide are offset by increases in oxides of nitrogen. The impact on the five indices is shown in Table C.30.

Comment: It should be remembered that this low cost of $3 is valid only when the minor tune-up is performed when the automobile is already at a garage—for example, if it is undergoing its annual inspection. Engine malfunctions are not corrected here, except for the two simple adjustments.

Annual Pollution Measurement with Diagnosis and Correction of High Emitters (TESTUNE); Change from Any Vehicle

Cost: In a test program, 50 percent of pre-1970 vehicles and less than 30 percent of 1970 and later vehicles were high emitters requiring maintenance.[23] When performed by independent garages, the average repairs cost $22.73 for pre-1968 vehicles, $23.37 for 1968–1969 vehicles, and $15.51 for 1970 vehicles. The simple idle emission test costs up to $1.00 per vehicle, assuming that the vehicle was in a test facility for other purposes, such as safety inspection.[24] All vehicles would be tested once and those failing would be tested again after repair, by the garage, then by the inspection facility. Fuel economy and other maintenance costs are assumed unchanged. If we ignore the cost of the initial test, the cost of the program can be attributed to the failing cars as shown in Table C.31.

23. A. J. Andreatch, J. C. Elston, and R. W. Lahey, "New Jersey REPAIR Project: Tune-up at Idle," *Journal of the Air Pollution Control Association*, 21, no. 12 (December 1971):757–763, Table II.
24. John N. Pattison, "Motor Vehicle Pollution Control News," *Journal of the Air Pollution Control Association*, 21, no. 10 (October 1971):656–658.

Table C.31. Cost (for failing vehicles)

Purchase ($)	Power Loss ($)	Total Capital ($/year)	Mainte- nance ($/mile)	Fuel ($/mile)	Total Variable ($/mile)	Total ($/mile)
—	—	—	0.00247	—	0.00247	0.00247 a
—	—	—	0.00254	—	0.00254	0.00254 b
—	—	—	0.00175	—	0.00175	0.00175 c

a. Pre-1968 vehicles, including repair and 2 tests.
b. 1968–1969 vehicles, including repair and 2 tests.
c. 1970 vehicles, including repair and 2 tests.

Table C.32. Effect

Pollutant	Crank- case HC	Evapo- ration HC	Total Exhaust HC	CO	NO_x	Pb	HC
Change from Pre-1968	—	—	—	−34	—	—	−32
Change from 1968–1969	—	—	—	−39	—	—	−11
Change from 1970	—	—	—	−41	—	—	− 8

Effect: Carbon monoxide and hydrocarbon emissions are both reduced for failing vehicles. When measured by the ACID mass emission method, the three vehicle groups experienced CO reduction of 42, 47, and 51 grams per mile and HC reduction of 6.3, 2.1, and 1.5 grams per mile. These are expressed as a percentage of base emissions in Table C.32. No mention is made of NO_x emissions, since they are not tested. It is likely, however, that they will be increased by repairs which reduce CO[25] but this is not included since no data are available.

These effects are observed immediately after repair. If we assume that emissions rise linearly over one year to pre-repair levels, the average reduction will be half this amount. This is reflected in the cost-effectiveness calculation given in Table C.33.

Comment: We disregarded the first measurement cost because it was

25. Paul B. Downing and Lytton Stoddard, "Benefit/Cost Analysis of Air Pollution Control Devices for Used Cars," Project Clean Air Research Project S-10, vol. 3, University of California at Riverside, September 1970.

Table C.33. Effect on Indices

Index	2 Net Percent	3 Minmax Percent	4 Percent Reduced	5 (2) with- out Lead	6 (3) with- out Lead	
Percentage	—16.5	0	—66	—22	0	Immediate
Change (%)						
(pre-1968)	— 8.2	0	—33	—11	0	Average
$\dfrac{\$/\text{mile}}{\text{Percent}} \times 10^4$	3.01	00	0.75	2.25	00	Pre-1968
	4.03	00	1.02	3.06	00	1968–1969
	2.87	00	0.70	2.13	00	1970

Table C.34. Cost (over BASE or PCV vehicle)

Type of Cost	Pur- chase ($)	Power Loss ($)	Total Capital ($/year)	Mainte- nance ($/mile)	Fuel ($/mile)	Total Variable ($/mile)	Total ($/mile)
User	35	0	8.31	0	0.00053	0.00053	0.00136
Resource	35	0	8.31	0	0.00036	0.00036	0.00119

small relative to repair cost. If a small proportion of vehicles need repair, however, the initial measurement is allocated to fewer improvements and may be significant. Furthermore, no value is placed on the inconvenience of taking the car for repair and returning to the test station; thus, these costs are probably understated. In any event they are as high as any new car program.

Used Car Kit (OLDKIT); Change from BASE or PCV Vehicle

Cost: The charge for the kit is $8 and, including installation, is no more than $35, according to local Chevrolet dealers. There is no loss in power or change in maintenance costs. Fuel economy is reduced by 2.25 percent, raising fuel costs by $0.00054 per mile to the user, and $0.00036 per mile resource. Removal of the vacuum advance reduces fuel economy, but the lean idle increases it, as does the correction of misfiring. In allocating these costs, it is recognized that eligible cars in 1972 will be at least five years old in most of the United States and seven years old in California. With a ten-year average life, the initial cost is spread over only five years, instead of ten. If the program does not require installation until sale, the average effective life is still shorter.

Effects: The effect of this kit, on the average, is to reduce hydrocarbons by 50 percent, carbon monoxide by 30 percent, and oxides of nitrogen by 30 percent.[26] This has the result, on a PCV car, shown in Table C.35. (See also Table C.36.)

Comment: It seems quite surprising that such relatively simple changes could so greatly affect total emissions. A large number of cars (over 200) were used, however, and the test procedures seem proper except that only hot cycles were considered and parts per million rather than grams per mile were measured. One source of error in practice would be that motorists who did not like the changed performance of a postkit car could easily undo the adjustments. Only field experience will show how serious a problem this may be, but the potential is ominous.

The changes of OLDKIT are more easily undone than the modifications which are made in new cars. There is thus a greater chance that unhappy

Table C.35. Effect (on PCV vehicle)

			Total				
	Crank-case	Evapo-ration	Exhaust				
Pollutant	HC	HC	HC	CO	NO_x	Pb	HC
Gross (gm/mile)	0	2.77	5.1	53.8	2.8	0.133	7.87
Change from PCV (gm/mile)	0	0	—5.1	—23.1	—1.2	0	—5.10
Change from PCV (% of BASE)	0	0	—50	—30	—30	0	—32

Table C.36. Effect on Indices (compared with BASE or PCV vehicle)

Index	2 Net Percent	3 Minmax Percent	4 Percent Reduced	5 (2) with-out Lead	6 (3) with-out Lead
Percentage change (%)	—23	0	—92	—30.7	—30
$\dfrac{\text{\$/mile}}{\text{Percent}} \times 10^4$	0.52	∞	0.13	0.39	0.40

26. G. W. Niepoth, G. P. Ransom, and J. H. Currie, "Exhaust Emission Control for Used Cars," Society of Automotive Engineers, Automotive Engineering Congress, January 1971, SAE Paper no. 710069.

owners will ask their mechanic to disconnect or nullify the kit. This factor would be used to discount the good cost effectiveness of the system.

That this kit reduces pollutants much more than the tune-up analyzed previously shows that the big gains in used cars are not in correcting malfunctions but in altering some basic engine parameters. This needs to be done just once, not annually, and is therefore relatively cheap.

If there is any error in the cost estimate, it lies in its failure to reflect the full reduction in gas mileage caused by the removal of vacuum spark advance. The estimate of 2.25 percent seems almost too small for the loss in fuel economy, as some technical papers suggest much larger effects from similar changes.

The figures shown are for a five-year-old vehicle with five years of life remaining. On older vehicles, the percentage cost rises, since the initial investment must be spread over fewer miles. Some examples are shown in Table C.37.

If mileage is accumulated more rapidly in early years than in later years of a car's life, the cost per mile and cost/net percent for OLDKIT will rise. Thus, there are several reasons why the device may not be as cost effective in practice as it appears here.

Table C.37. Costs for Older Vehicles

Vehicle Age	Capital Cost/Year	Total Cost/Mile	Cost/Net Percent
7	$13.09	0.00167	0.73
9	$35.00	0.00386	1.68

Index